Towelsail Yawls

The Lobsterboats of Heir Island and Roaringwater Bay

Cormac Levis

GHP
Galley Head Press

First published in 2002 by
Galley Head Press
Dunowen
Ardfield
Co. Cork
(023) 40881

British Library Cataloguing in Publication Data
A CIP catalogue record for this book is available from the British Library

ISBN: 0 9542159 0 7

Typesetting: *Dominic Carroll, Co. Cork*
Printing: *ColourBooks, Dublin*

Cormac Levis is a native of Ballydehob, a village situated on an inlet of Roaringwater Bay in West Cork. For generations, his family worked sailing sand-dredgers out of this small, tidal harbour. In the late 1970s and early 1980s, he fished extensively off the south coast, at a time when many veterans of the towelsail yawls were still going to sea.

Cormac is one of a nationwide team of maritime researchers involved in Traditional Boats of Ireland, a co-operative project for the publication of a major study on traditional Irish boats. He has also written comprehensively on the West Cork sandboats for the *Mizen Journal* and other publications. He has lectured on both the lobsterboats and sandboats at the Glandore Classic Boat Summer School and at various historical-society meetings, and has contributed to radio programmes on the subject. He is a founding committee member of the Ilen River Cruising Club.

Cormac Levis is a secondary-school teacher at St Fachtna's De La Salle, Skibbereen.

The author, Cormac Levis (*right*), in conversation with Jack Pyburn, grandson of Heir Island boatbuilder, Richard Pyburn, and a veteran of the towelsail yawls.
Photograph: Christine Thery

For Saoirse
13 August–19 October 1998

Come away, O human child!
To the waters and the wild

CONTENTS

ACKNOWLEDGEMENTS

The principal source for this book has been the living folklore of Roaringwater Bay and West Cork. My sincere thanks to the following for their memories, stories and hospitality, and may God rest those among them who have died since I began research for this book in 1996:

Jack Pyburn, Heir Island; Neily O'Donovan (deceased), Heir Island; John and Kathleen Harte (both deceased), Kilkilleen and originally of Heir Island; John and Bridie Fitzgerald, Skibbereen and originally of Heir Island; Willie O'Regan, Whitehall and originally of West Skeam Island; Kate O'Driscoll, Kilkilleen and originally of Heir Island; Denis O'Driscoll (deceased), Kilkilleen and originally of Cape Clear; Denis O'Sullivan, Colla, Schull; Dan Griffin (deceased), Schull; Anthony Griffin, Colla, Schull; Jerry O'Mahony, Crookhaven; James O'Regan (deceased), Crewe, Schull and originally of Long Island; Jimmy O'Reilly (deceased), Schull; George Bushe, Crosshaven; Seán Harte, Courtmacsherry and Heir Island; Mick O'Brien (deceased), Clashmore, Ballydehob; Connie McCarthy, Ballycummisk, Ballydehob and originally of Horse Island; Tim O'Sullivan, Long Island; Tom Supple, Crookhaven; Bobby Allen, Toormore; Christy Burke, Schull and originally of Heir Island; Pat O'Driscoll, Sherkin Island; Patrick O'Regan, Schull; Micheál Ó Dálaigh (deceased), Turk Head and originally of Cape Clear; Michael Minihane, Garrettstown and originally of Heir Island; Agnes O'Donovan (deceased), Kilkilleen; Gene O'Neill, Oldcourt; Cliffy Minihane, Cunnamore; Jerry Minihane, Cunnamore; Breda Harte, Laheratanavally; Seán Barnett (deceased), Schull; John O'Neill, Church Cross; Michael Whooley, Kilkilleen; Denis Minihane, Wiltshire, England and originally of Cunnamore; John Joe O'Sullivan, Gubbeen, Schull and originally of Long Island; Jim McGrath, Dunmanus; Mack O'Donoghue, Cape Clear; Seán Connolly, Oldcourt; John O'Callaghan, South Schull; Tony Welland, Monkstown; Clayton Love, jnr., Cork; Brian Love, Dublin; Denis O'Donoghue, Caherdaniel, Bantry; Donal Dinan, Cork; Michael Daly, Cape Clear; Michael John O'Neill, Whitehall and originally of Heir Island; the Murphy family, Cunnamore and originally of Heir

Island; Dermot McSweeney, Baltimore; Ralph Skinner, Rathmore, Baltimore; Mary Dwyer, Collatrum and originally of East Skeam Island; John Maloney, Rineen, Castlehaven; John Maloney, Union Hall; Cora Crowley, Castletownbere; Paddy O'Regan, Gortnamona, Schull; Lizzy Minihane, Heir Island; Dan McCarthy, Heir Island; Jim Downey, Toormore; Thomas Nash, Cobh; John Hunt, Bantry; Peter Cullen, Cork; Eddie English, Cobh; Denis O'Sullivan, Schull and originally of Long Island; Francie Healy, Old Head of Kinsale; Eddie Pyburn, Schull and originally of Heir Island; John James Goggin, Schull; David Finn (deceased), Gurteenroe; Seamus Davis (deceased), Poulnacallee; Margaret O'Donovan, Skibbereen and originally of Heir Island; Eugene Gillan, ex-Principal Light-keeper, Old Head of Kinsale; James Cadogan, Ardagh; Bridie Cadogan (deceased), Poulnacallee; Derek Harte, Schull; Timmy Joe Whooley, Fasagh.

I am indebted to the following for old photographs of the lobster-boats: Agnes O'Donovan, Kilkilleen; Margaret O'Donovan, Skibbereen; Diarmuid Minihane, Skibbereen; May O'Regan, London; John James Goggin, Schull; Richard Bushe, Baltimore; Michael Casey, Union Hall; Donal Lynch, Cork; Patrick O'Regan, Schull; Tony Welland, Monkstown; Michael John O'Neill, Whitehall.

Thanks to Kevin O'Farrell, photographer, Heir Island; Nigel Towse, Sherkin Island; Mary Jordan and Colm Ó Cuileannáin, Baltimore; Norcott Roberts, Monkstown; and Christine Thery, Heir Island, for recently taken photographs.

Photographs from the Lawrence Collection (*photo refs*: C5191, R4862, R10206) have been reproduced with the permission of the National Library of Ireland; photographs from the Fr Browne Collection with permission from the Irish Picture Library, Dublin; photograph from the Greene Collection, copyright of the National Museums and Galleries of Northern Ireland, Ulster Folk and Transport Museum (*photo ref*: WAG 2359); photograph on p. 89 with the permission of Giles Norman Photography. Thanks to Terence Spillane, Dublin, and the National Library of Ireland for permission to reproduce photographs from the Spillane Collection. A special thanks to Darina Tully, Dalkey, Co. Dublin for sourcing the photographs from the Fr Browne, Lawrence and Greene collections.

Drawings and quotations from the Ó Síothcháin manuscripts have been reproduced with the kind permission of the head of the Department of Irish Folklore, University College, Dublin. Thanks to staff member, Criostóir MacCárthaigh, for his invaluable assistance in sourcing this material and also the photographs from the Spillane Collection.

I am indebted to Ted O'Driscoll, Baltimore for recording and drawing up the lines plan of the *Hanorah*, to Russell Barnett, Dunmanway for illustrating my description of pot-making, and to Richard Power, Bank, Castletownbere for permission to reproduce the drawing of a lobsterboat at Bank Harbour.

Thanks to Liam and John Hegarty, Nigel Towse and Fachtna O'Sullivan of Hegarty's Boatyard, Oldcourt for clarification of certain aspects of traditional boatbuilding, and to Cian Ó Sé, Cork for etymological research.

I owe a debt of gratitude to Tim Cadogan, not only for his professional assistance as archivist in Cork County Library, but also for the provision of relevant newspaper reports sourced during the course of his own personal research in maritime history. The assistance of the staff of the reading rooms at the National Library of Ireland is also appreciated. Thanks to Liam O'Regan, editor of the *Southern Star* newspaper, for access to its archives and those of the *Eagle*, and to staff member John Hamilton for his assistance. I am grateful to Michael Carey, Registrar of Shipping for the Port of Registry of Skibbereen, both for facilitating access to the *Register of Sea Fishing Boats* and for consulting the records of certain vessels on my behalf. Thanks also to Seamus Ryan, Superintendent Registrar of Births, Marriages and Deaths at Skibbereen, and to his assistant, Eileen O'Sullivan.

My sister, Mary Margaret Fitzgerald, has given generously of her time and expertise in word-processing, and O'Donovan, Murphy & Partners, Bantry kindly facilitated me with the use of their computers, while Jennifer Corcoran, Ballydehob was kind enough to proof-read for me. Thanks to Liam Blake, photographer, Wicklow and Ger Garland, Dublin for technical advice, and to Dominic Carroll, Ardfield not only for his excellent work in typesetting and final proof-reading but also for his advice and support in the final stages of the production of this book.

I wish to acknowledge my involvement in the Traditional Boats of Ireland project as the catalyst that transformed my interest in the towelsail yawls into an obsession, and, in particular, the role of Donal Lynch and Ted O'Driscoll who first encouraged me to write about these boats. The advice and personal encouragement of Kevin and Jane O'Farrell of Heir Island has also been deeply appreciated.

To my late father, Jim Levis, and my late uncle, Sim Levis, I owe my inherited love of boats and the sea, and to the patience, understanding and support of my wife Cearma, I owe the pleasure of being able to indulge in it.

Cormac Levis
December 2001

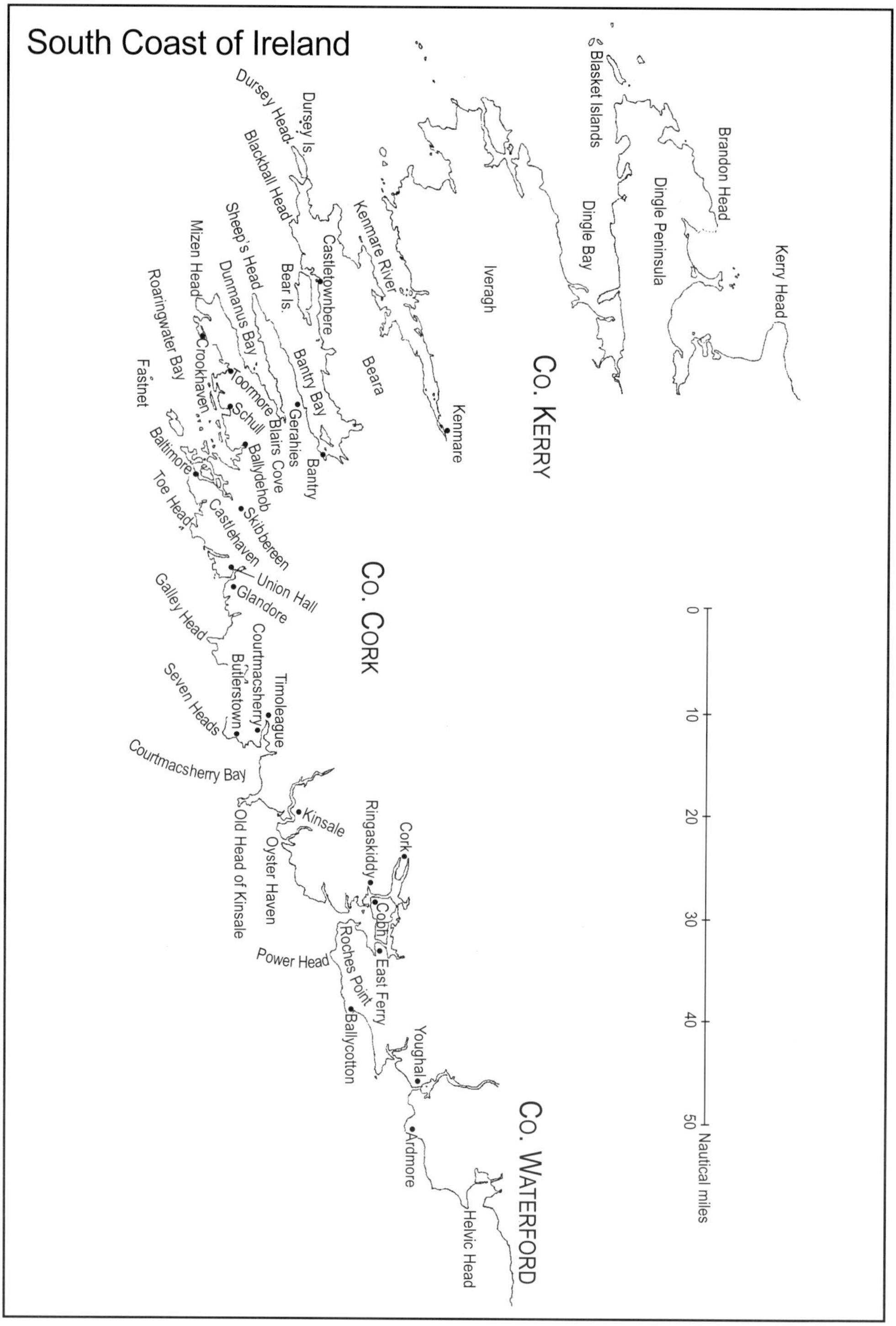
South Coast of Ireland
Co. Kerry
Co. Cork
Co. Waterford
Blasket Islands
Brandon Head
Dingle Peninsula
Kerry Head
Dingle Bay
Iveragh
Dursey Head
Dursey Is
Blackball Head
Kenmare River
Castletownbere
Bear Is.
Beara
Kenmare
Sheep's Head
Bantry Bay
Bantry
Mizen Head
Dunmanus Bay
Gerahies
Blairs Cove
Roaringwater Bay
Crookhaven
Toormore
Schull
Ballydehob
Fastnet
Baltimore
Skibbereen
Toe Head
Castlehaven
Union Hall
Glandore
Galley Head
Courtmacsherry
Butlerstown
Timoleague
Seven Heads
Courtmacsherry Bay
Kinsale
Old Head of Kinsale
Oyster Haven
Ringaskiddy
Cork
Cobh
East Ferry
Roches Point
Power Head
Ballycotton
Youghal
Ardmore
Helvic Head
0
10
20
30
40
50
Nautical miles

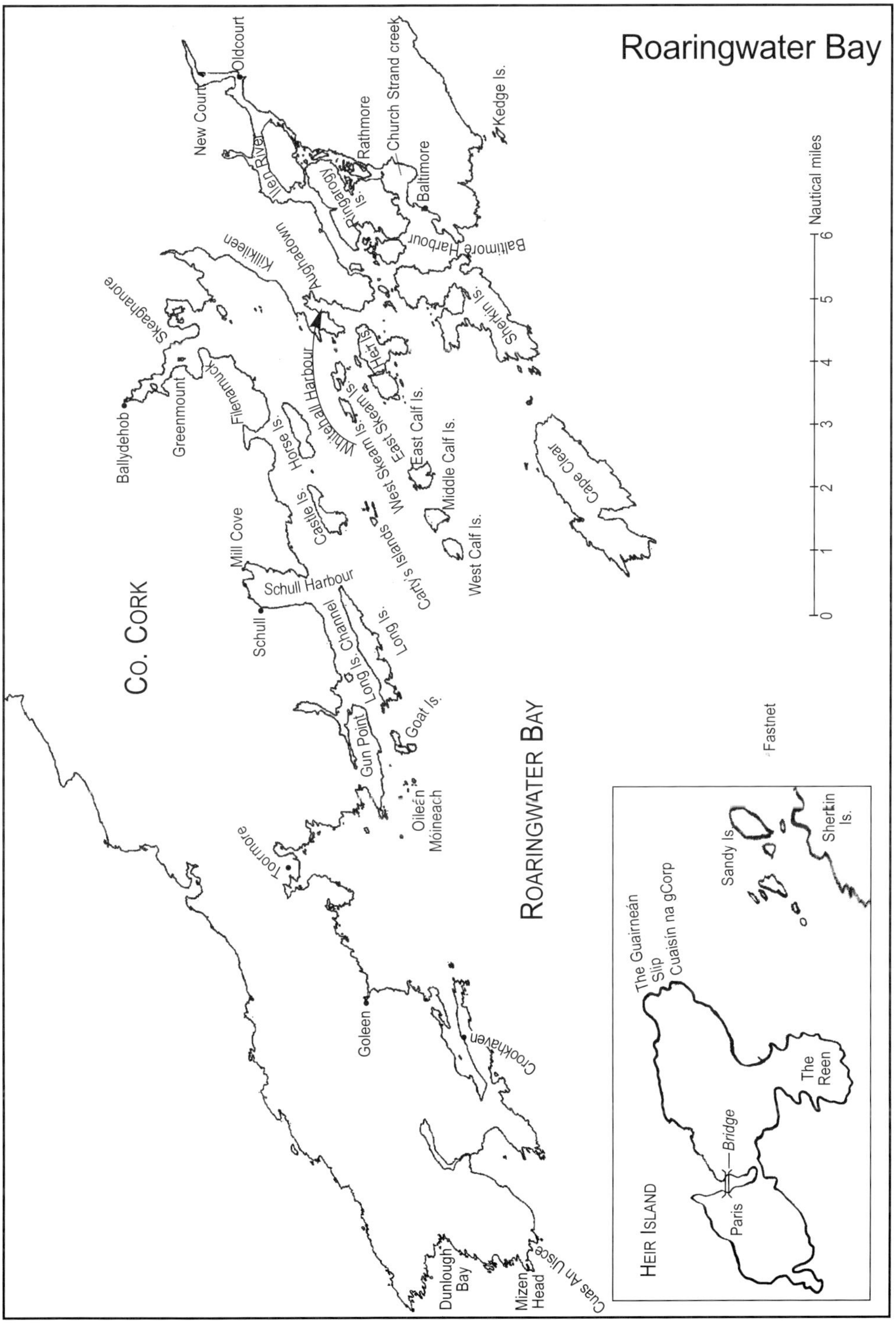
Roaringwater Bay
Co. Cork
Roaringwater Bay
New Court
Oldcourt
Ilen River
Church Strand creek
Rathmore
Kedge Is.
Ringarogy Is.
Baltimore
Baltimore Harbour
Kilkilleen
Aughadown
Skeaghanore
Sherkin Is.
Heir Is.
Whitehall Harbour
Fileenamuck
Greenmount
Ballydehob
Horse Is.
West Skeam Is.
East Skeam Is.
East Calf Is.
Middle Calf Is.
West Calf Is.
Cape Clear
Castle Is.
Carty's Islands
Mill Cove
Schull Harbour
Schull
Long Is. Channel
Long Is.
Goat Is.
Gun Point
Oileán Móineach
Toormore
Fastnet
Goleen
Crookhaven
Dunlough Bay
Mizen Head
Cuas An Uisce
Nautical miles
0
1
2
3
4
5
6
Heir Island
Paris
Bridge
The Reen
The Guairneán Slip
Cuaisín na gCorp
Sandy Is.
Sherkin Is.

1 · An Untold Story

Hidden away in a quiet corner of a hushed and sheltered inlet near Schull Harbour was, until recent years, a sad and dismal little wreck. Pinned firmly to the strand by a fallen tree, her dark, slime-covered bilge and the blotchy, peeling paint of her topsides blended chameleon-like with the muddy foreshore and the thick, overhanging vegetation growing on the bank behind her. If by chance she attracted a first glance, she was rarely accorded the dignity of a second, except perhaps by the few who saw beyond her battered nameless bow to the *Hanorah* of a hundred years earlier.

In those more glorious days of the 1890s, the *Hanorah* was part of a large fleet of boats that set out under sail and oar each year from Roaringwater Bay to fish lobsters and crayfish all along the shore from Dursey Head to Ardmore in Co. Waterford. Once a familiar sight in every bay, inlet and harbour on the south coast, the lobsterboats of Roaringwater Bay were among the best-known and most distinctive fishing boats of the region. In the late 1940s and

The dismal state of the *Hanorah* in 1998. A hundred years earlier, she was a beautiful and graceful little boat, one of several that set out every year from Roaringwater Bay on long fishing trips. *Photograph: Cormac Levis*

early 1950s, many of them were fitted with engines and modernised before gradually falling into disuse in the following decades.

The untold story of these boats has been a missing chapter in Ireland's maritime history. It is a story, however, that has survived in the hearts and minds of those who were themselves part of it. Away from home for periods of up to seven weeks' duration and constantly on the move, the fishermen led a semi-nomadic lifestyle, working, cooking, eating and sleeping in their small, open boats. Many years later, some of these men still recall not only their own youthful experiences on the long fishing trips in the lobsterboats, but also the stories of the generations who went before them. For those who care to listen, romantic images of a carefree life on the boundless ocean gradually dissolve and merge with vivid accounts of the harsh realities of everyday life – and death – to give an all the more fascinating, if at times melancholy, insight into a way of life that has gone forever.

2 · A Unique Fleet

In the final decades of the nineteenth century, the fishing industry on the south Cork coast was a vibrant one. By 1890, Baltimore had become one of the most important ports in Ireland for the mackerel fishery, with large volumes of mackerel being landed there by a huge fleet of mackerel drifters from the Isle of Man, England, Scotland, France and Ireland. Smaller, but very significant, amounts of mackerel were also landed at Castletownbere, Crookhaven and Schull. The total number of mackerel drifters fishing off the south coast in the spring of 1881 was 737.[1] Cape Clear, which had a fleet of these large mackerel drifters, also had a fleet of over fifty small boats which fished mackerel, with moored nets, when the shoals came close inshore in autumn. On the Beara Peninsula and in Dunmanus Bay, sein fishing, an age-old method once employed to catch pilchards, was now being applied to the mackerel fishery with great success. At Gerahies, on Sheep's Head, the herring fishery was very actively pursued. All along the south Cork coast, but especially in areas like Myross, Glandore and Courtmacsherry, a fleet of small boats travelled incredible distances out to sea to catch a huge variety of fish on long-lines and handlines.

In the midst of all of this activity in the fishing industry, however, lobsters remained a largely under-exploited species. This is reflected in the 1892 *Baseline Reports* of the Congested Districts Board which gives us an indication of the generally neglected state of the lobster fishery on the south coast.[2] There was, however, as we shall see, one small, highly localised community of fishermen who were unique on this coast in their dedicated pursuit of the lobster.

To the north of Roaringwater Bay, there was a total of fifty-five fishing boats based on the sparsely populated Sheep's Head Peninsula, but none of them engaged in lobster fishing. When the Congested Districts Board inspector visited Castletownbere, there were 174 mackerel drifters landing there, at least 162 of which were from other ports, and a total of 194 boats of various classes were registered as being owned locally. Yet on the entire Beara Peninsula, only twenty boats were 'occasionally employed' in lobster fishing, and it did 'not seem to be pursued in a regular manner'. In the Kenmare District, there was one trawler and a total of eight other boats, none of which engaged in lobstering. To the east of Roaringwater Bay, there were no lobsterboats in Castlehaven. It is known that there was a certain amount of lobster fishing carried on in Glandore, which was outside of the Congested Districts area; but Seán Ó hAo, a local fisherman born in 1861, gave the impression that it was not an important fishery there.[3] Further east, there were only four boats lobstering in Butlerstown. In Courtmacsherry, where there was a total of thirty-six other boats, mostly engaged in line fishing, there was only one lobsterboat.

Even in the environs of Roaringwater Bay itself, there were only thirteen lobster boats in the Congested District of Schull, which encompassed the entire Mizen Peninsula (including the Dunmanus Bay side) and the islands of West Calf, Middle Calf, Horse Island, Castle Island, Long Island and Goat Island. There was a total of 113 fishing boats registered as being owned in the district. It is only when we come to the Congested District of Baltimore, which included the remaining seven inhabited islands in the bay, that we find a significant number of lobsterboats. There were 211 fishing boats registered as being owned in this district, and there were forty lobsterboats based there. What was really remarkable about the large number of lobsterboats present there, however, was the fact that the entire fleet was concentrated in just one of the four electoral divisions that made up the district, namely that of Aughadown South. Even within that area itself, the oral tradition narrows down the base of the fleet even further, placing its nucleus on Heir Island, where approximately three-quarters of the boats were owned. The remaining lobsterboats were on Heir Island's near neighbours of East Skeam and West Skeam, and on mainland Aughadown. According to the 1891 census, Heir Island (381 acres) was home to forty-five families, which accounted for a total population of 325. East Skeam (fifty acres) had three families, namely the Barrys, Cadogans and Desmonds, giving a total population of twenty. On West Skeam (thirty-one acres) the two O'Regan families accounted for its total population of sixteen. Jim Barry and Tim Desmond of East Skeam each had a lobsterboat. By the early 1900s at least, Rickie O'Regan's family on West Skeam had two.

The Congested Districts *Baseline Reports* do not quite cover the entire south Cork coast, but there seems to be no historical evidence or folk-memory of any significant numbers of lobsterboats concentrated in any other part of this coastline. In 1892 then, the year before the *Hanorah* was registered, we find a well-established fleet of boats concentrated in a corner of Roaringwater Bay, which was unique on the south coast as a relatively large, distinct, cohesive, highly localised fleet specialising in lobster fishing. It is important to note that this fleet was already in place before the Congested Districts Board started to provide funding to fishermen for the purchase of boats, the *Baseline Reports* being an account of the economic situation on the ground prior to the intervention of the Board.[4] What are sometimes perceived as local fleets of fishing boats indigenous to certain areas were, in fact, introduced by the Congested Districts Board from 1896 on. This fleet also had another claim to distinction. Lobstering was primarily an inshore fishery usually pursued in local waters in four-oar rowing-boats of twenty feet to twenty-four feet in length. The boats of Heir Island and its neighbourhood, on the other hand, were gaff-rigged sailing boats with a crew of three rather than four,[5] and ranged out over huge distances to exploit the under-fished waters to the east and west of them. To many people, especially those living outside of the Roaringwater Bay area, they were known collectively as the 'Heir Island lobsterboats'. Locally, they were usually known simply as the 'lobsterboats', but they were also often referred to as the 'towel-sail yawls', the meaning of which will presently become evident.

Unlike the Heir Island fleet, the small number of lobsterboats in that area of the bay covered by the Schull Congested District were basically rowing-boats with a dipping lugsail. In the 1890s and early 1900s, however, many of the lobstermen of this area had slightly deeper, larger boats built with a gaff rig similar to those of Heir Island. The number of the crew was reduced from four to three, and they started to make lobster fishing trips 'up north' as far as Dursey Head. One of the first fishermen to take to lobster fishing at this side of the bay was Charlie O'Regan of West Calf. He is believed to have had a lobsterboat from at least as early as the 1880s. In the early 1900s, his sons Dan and Patrick had gaff-rigged lobsterboats built. The two Scully families of the Middle Calf had done likewise by this time. On Long Island,[6] an increasing number of fishermen were taking to lobstering, and although some remained faithful to the old dipping lugsail, many of them also changed to the gaff rig. Some Cape Clear men also took to lobstering in the early 1900s, but the mackerel fishery was by far the dominant fishery there, and lobstering was not pursued with the same long-term commitment and single-mindedness.

This fleet of sailing lobsterboats then, from Heir Island, East Skeam, West Skeam, Aughadown, Middle Calf, West Calf and Long Island, were the towelsail yawls of Roaringwater Bay that dominated the lobster fishery of the south coast down to the 1950s. In the following text, they are more frequently referred to by the popular term 'lobsterboats'.

3 · The Boats and Their Builders

According to tradition, a ship's carpenter on a vessel trading between England and Oldcourt over 200 years ago fell in love with a local girl whom he married and settled down with on the banks of the Ilen River opposite Oldcourt. His name was Pyburn, and he had four sons, all of whom he trained as boatbuilders. One of these sons, Christopher Pyburn, was a frequent visitor to Heir Island where he repaired boats for the islanders. Eventually, he settled down on the island, marrying a woman of the McCarthy Rábach family who, with her brother, had moved to the island from their mainland home at Filenamuck, Ballydehob. Christopher and his son Richard built many lobsterboats on Heir Island in the second half of the nineteenth century, and Richard continued to build them into the early years of the twentieth century.

Christopher Pyburn's brothers settled in other localities. One of them set up business on the shores of Dunmanus Bay near Durrus, and another, Tom, at Toormore, Goleen. Tom's sons, Dick and Tom, continued to build boats at Toormore down to the late 1930s. Another son, Jack, set up business at Crookhaven around 1904. It was these Pyburns of Toormore and Crookhaven who built most of the lobsterboats of Long Island and the Calf Islands.

Besides the Pyburns, of course, there were several other boatbuilding families producing lobsterboats. From the 1860s on, Willie Skinner of Rathmore, Baltimore built lobsterboats and passed on the craft to his sons Willie and Henry. It was Henry who established Skinner's Boatyard in Baltimore village early in the twentieth century, while his brother continued to build at Rathmore. Flor O'Neill of Sherkin Island built several lobsterboats, especially in the early 1900s after he had moved to Baltimore. At The Cove, to the south-west of Baltimore village, a small community of boatbuilders, comprised of the Minihane, O'Driscoll and Bushe families, built lobsterboats up to the 1890s.

Each boatbuilder had his own particular style, and a discerning eye could distinguish between the boats of various builders; nevertheless, they were all very similar in basic design, having developed from the old spritsail boats that were once, alongside the larger hookers, the typical small fishing boat of Roaringwater Bay.[7]

Like all traditional West Cork boats, the lobsterboats were of carvel construction. They were lightly built with a moderate sheer, low freeboard, narrow beam, mildly raked transom and a stem-post that curved smoothly from the keel to a near-upright position, yet never quite straightened. These features gave them a graceful, streamlined appearance.

Generally, the lobsterboats ranged in size from twenty-four feet to twenty-eight feet in overall length, although some, especially in Long Island, were as small as twenty-two feet. A typical lobster-boat built by Richard Pyburn of Heir Island in the 1890s would measure approximately twenty-six feet in overall length, six and a half feet in breadth and three to three and a half feet in registered depth. Just large enough to carry a substantial number of lobsterpots and all the paraphernalia required for a long fishing trip, they were small enough to manoeuvre in the dangerous inshore waters and narrow inlets they frequently fished. They were deep enough to sail, but with their light build, low freeboard and narrow beam, they could also be easily rowed.

The lobsterboats were usually planked with pitch pine, although red deal and white deal were also used. Locally felled oak, or sometimes elm, was used for the frames. Some boatbuilders judged the suitability of an elm tree for boatbuilding purposes by chewing its leaves. Shipwrecked cargoes of timber often provided the planking, and New Court woods, on the banks of the Ilen River, was a favourite source of oak and elm. Large trees inclining out over the river before sweeping skywards produced the natural 'bends' eagerly sought after by boatbuilders. New Court was particularly convenient for Richard Pyburn, as the timber could be transported by boat directly from the riverbank to Heir Island.

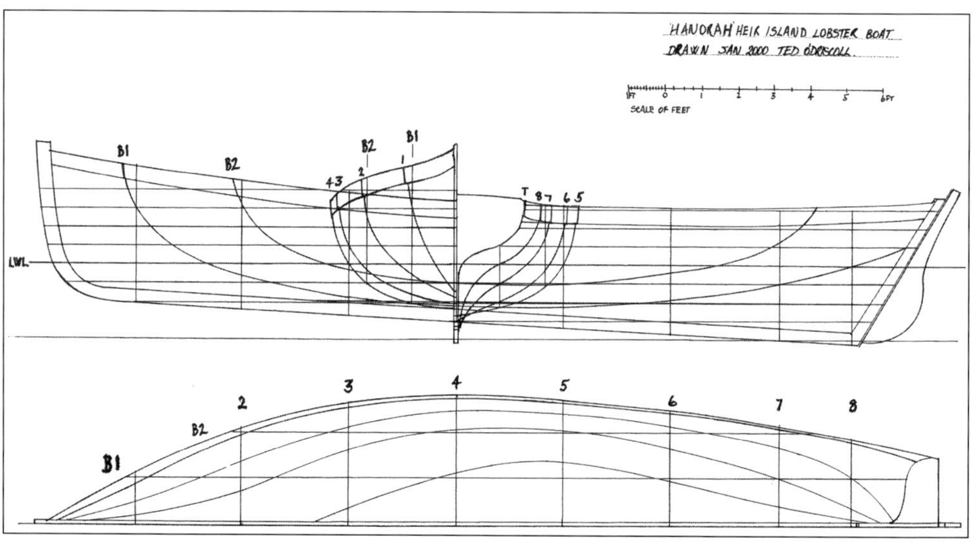

The graceful lines of the *Hanorah*, built by Richard Pyburn, Heir Island in 1893 and drawn by Ted O'Driscoll, Baltimore in 2000.

The timber was cut up into planks and slabs by means of a saw-pit. In large boatyards, saw-pits were long, narrow, stone-lined pits about eight feet deep. The deal or log that was to be cut up was placed lengthwise over the pit so that two men, one standing on the log and one in the pit, could cut it using a two-man saw known as a pit-saw. The man standing in the pit had the hardest share of the work as the saw cut was made on the downward stroke. The man on top was responsible for guiding the blade. Locally, saw-pits usually took the form of a raised, wooden platform rather than a pit. Nevertheless, they were still referred to as 'saw-pits'. On Heir Island, Richard Pyburn employed two professional pit-men from Skibbereen, known as 'croppies', to saw up his timber. They also felled the trees at New Court.

Small boatbuilders built their boats in the open. Richard Pyburn built his in the field near his house known as the Bán. An interesting feature of the construction of these boats was the use of traditional oak treenails, rather than the iron bolts or drifts used in later years. The entire backbone of the boat, consisting of keel, stem-post, apron,[8] sternpost and deadwoods, was held together by treenails. The beautifully shaped, 'wine glass' transom was spiked to the inside of the sternpost, giving rise to the traditional, external sternpost on which the rudder, or 'steer',[9] would eventually hang. The frames were usually three-piece single frames of grown oak consisting of two top timbers that extended in one piece right down to the ground timber, or floor. No hog[10] was used and the frames were treenailed directly to the keel. Each frame and the transom were notched so that the top of the plank below the sheer strake, known as the 'levelling strake', sat in flush with the outside edge of the frames. This allowed the sheer strake to overlap the levelling strake and extend down over it in excess of one inch. The levelling strake was always the first plank fitted, and the bottom of the sheer strake was fastened to it.

The notching of the frames was carried out after the frames were put in place and the sheer line and the position of the levelling strake had been determined. A short wedge of timber was removed from the outside face of each frame along the intended line of the top edge of the levelling strake. This was done by making a vertical saw cut, from below up, on each frame and then severing the top of the wedge from the frame with two sharp, accurate blows of an adze.[11] It is said that a skilled shipwright could carry out this procedure with seemingly effortless speed and accuracy, moving quickly from frame to frame.

Skillful use of the adze was a source of pride to a good shipwright, and it was said that a man who could not split a pencil line with it was not fit to use it. One boatbuilder demonstrated his accuracy with this tool by removing the brim of the top hat of a client who had found fault with his workmanship. His point was

Photograph from the Lawrence Collection showing a lobsterboat under construction at Baltimore in the 1890s. The notches in the frames to take the top of the levelling strake, clearly visible in this photograph, were a notable characteristic of traditional West Cork boatbuilding. Among the boatbuilding tools visible is an adze, standing near the boat, and in the foreground, a caulking mallet. The planked roof in the background was a common feature of buildings associated with West Cork boatyards.

all the more forcefully made by the fact that the client was wearing the hat at the time.

Bolted to the outside of the hull, up against the overlapping edge of the sheer strake and centred on the position of the main thwart, were two legwales, one on either side. They were four to six feet in length, up to three inches wide and two inches in height, and tapered in towards the hull at either end. They were referred to as 'bends', and were mortised to allow the chainplates for the shrouds to pass down inside of them. The legs themselves had a deep notch cut into them some inches from the top. A hole was drilled from the top of the leg vertically down through its centre until it reached this notch. The legs, of course, were used to keep the boat sitting in an upright position when dried out on the shore. When in use, a rope was threaded through the notch and up through the hole, coming out through the top of the leg. A knot in the end of the rope at the notch prevented it pulling through. The end that protruded from the top of the leg was taken up through a hole in the legwale, in over the gunnel, pulled tight and made fast inside the boat keeping the leg jammed firmly under the legwale.[12] In later years, the legs were bolted on.

The use of grown oak to avoid doubling in frame construction[13] was a measure primarily aimed at making the boat as light as possible. The overlapping sheer strake gave rigidity to what was essentially a lightly constructed boat, and helped to prevent longitudinal twisting.

The advantages of lightness in a boat that had to be rowed and sailed long distances, and sometimes hauled up on the shore with

The adze, an ancient boat-building tool, still used by traditional boatbuilders like Fachtna O'Sullivan in Hegarty's yard at Oldcourt.
Photograph: Kevin O'Farrell

limited manpower, are obvious. Economic factors, however, also probably played a large part in the design and development of these boats. The less timber used in the construction of the boat meant not only a lighter boat, but also that the boat was more economical to produce. The importance of economic considerations for prospective boat owners is illustrated by the fact that boatbuilders offered their clients a choice of qualities in the oak used in the construction of their boat, with the best heartwood making the difference of £1 in the overall cost of an average-sized lobsterboat in the 1890s. At that time, a typical lobsterboat built on Heir Island by Richard Pyburn, including mast, gaff and oars, cost from £14 to £16.

The lobsterboats were completely open boats except for a very short foredeck, about two feet in length, under which there was a locker. They had two thwarts, referred to by the local term 'thaufts' (or 'thafts').[14] In a twenty-seven-foot boat, the main thauft was about nine feet from the stem. The aft thauft, known as the 'hind thauft', was about six feet from the stern. The main thauft was also known to some as the '*béal tuille* thauft'. The area between the two thaufts was referred to as the 'holt'. The area forward of the main thauft, usually referred to as the 'fore-sheets', was sometimes

Victory II, (*below*) owned by Michael Coughlan of Kilkilleen, Aughadown, and built at Baltimore in 1910. When dried out, the boats were supported on legs attached to legwales, known as 'bends'. The boats were tarred from the waterline down but most had painted topsides with the overlapping top strake given a different colour to the rest of the hull. Photograph taken about 1940.

referred to as the '*béal tuille* sheets'. The stern-sheets, which is the area abaft the after thauft, was often referred to as the 'hind-sheets'. In early times, the holt was seldom, if ever, floored. Instead, the ballast, which was stone, was levelled off with small stones packed between the larger ones to give an even surface which could easily be walked upon. An easily removable, short plank in the floor of the stern-sheets gave access to the 'well', the deepest part of the bilge, for the purpose of bailing. The preferred ballast was heavy, hard, black stones which were hand-picked along the shore. In the late 1930s and 1940s, most of the boats were fully floored, and iron ballast, salvaged from the wrecks of large, old, abandoned mackerel drifters, was frequently used.

During the 1870s, a boomless, gaff mainsail and a foresail replaced the original spritsail rig of the lobsterboats.[15] The mast was positioned against the aft side of the main thauft, and the thauft knees

During the 1870s, the original spritsail rig of the lobsterboats was replaced by a boomless, gaff mainsail and a staysail as seen in this photograph of the *Mary Hannah*, taken in the early 1930s. On board is her owner, Jeremiah McCarthy of Paris, Heir Island at the stern; and from left to right, his brother Michael, nephew Cornelius Harte and brother-in-law Timothy Harte. On shore is his mother Mary McCarthy.

extended in as far as the mast, providing further support at either side. The cleats for belaying the halyards took the form of short iron bars passing down through the thauft. There were two wire shrouds at either side, which had thimbles spliced in their ends and were set up to the chainplates with rope lanyards. The luff of the foresail was loosely laced to the forestay, and the luff of the mainsail was also quite loosely laced to the mast to avoid jamming when hoisting. A horse for the mainsheet spanned the width of the transom. From the early 1900s at least, some lobsterboats also carried a jib, and a few fished with almost a full cutter rig of mainsail, staysail, jib and topsail. Even boats that usually fished with just the mainsail and foresail, frequently carried a topsail and jib for regattas, which were keenly contested. In this rig, the bowsprit passed through a gammon iron on the port side of the stemhead. In most boats, the inboard end of the bowsprit was simply lashed down with a rope. In some boats, however, it was secured by means of an iron fitting which was bolted to the short foredeck. It is not clear how this fitting was designed, but it involved inserting an iron pin to lock it in position around the heel of the bowsprit. The jib was set by means of a traveller ring and outhaul. The outhaul fall was passed around a thumb cleat low down on the starboard side of the stem, before being brought inboard and heaved up tightly to act as a bobstay. Quite often, the thumb cleat was no more than a wooden dowel protruding from a hole drilled in the

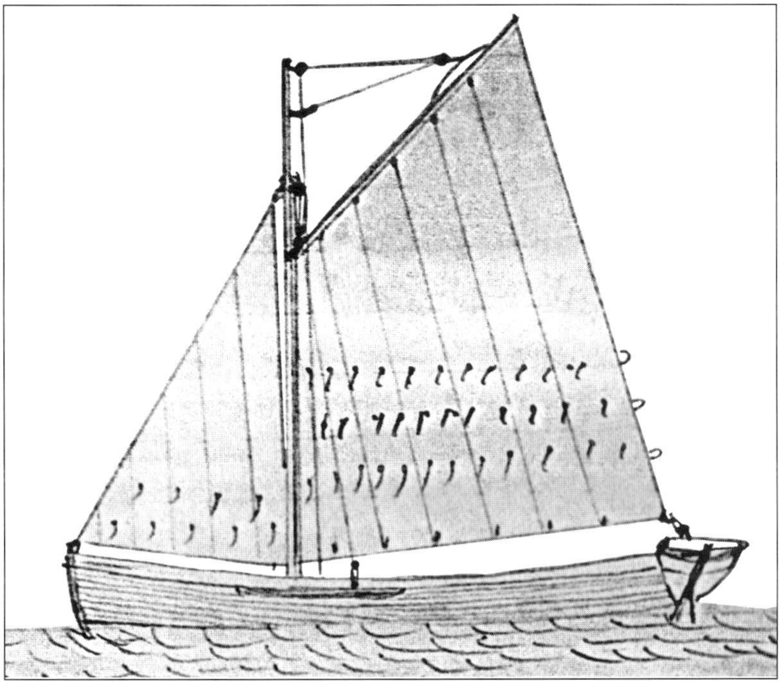

Sail plan of a Roaringwater Bay gaff-rigged lobsterboat, as drawn by Ciarán Ó Síothcháin, Cape Clear in 1952 (IFC 1320, p. 490), a few years after the rig had been made more or less redundant by the introduction of engine power.

A mackerel yawl from the early twentieth century drawn from memory by Ciarán Ó Síothcháin of Cape Clear in 1958 (IFC 1517, p. 59). From the early 1900s at least, while the mainsail and foresail remained the usual working rig of most lobsterboats, a few also carried a jib and topsail, a rig which would have closely resembled the rig depicted here.
The lobsterboats, however, did not have a boom on the mainsail, and the outhaul of the jib was returned through a thumb cleat low down on the starboard side of the stem-post to form a bobstay. In later years at least, some boats had a triangular topsail set on a standing yard rather than the four-cornered one seen here.

Most boats usually fished with just the mainsail and foresail, but even these would frequently set a jib and topsail for regattas, which were keenly contested. This photograph of a lobsterboat at Schull Regatta was taken around 1925. Unfortunately, the topsail is not very clear. The fall of the jib outhaul, which acted as a bobstay, is faintly visible as it returns inboard through a thumb cleat low down on the stem-post. The very short foredeck is also visible.

Schull Regatta, 1935. On the left of the photograph, the lobsterboats can be seen clustered at the stern of the committee boat. The silhouette of some topsails can be made out.

(above) George Bushe's interpretation of the regatta rig as seen on his model of a lobsterboat. George's ancestors built lobsterboats at The Cove, Baltimore.
Photograph: Kevin O'Farrell

(right) The *Mary Ellen*, owned by Denis Minihane of Conomar, Aughadown. With close observation, one can make out the jib that this boat always carried as a working sail.

The early Long Island boats were basically rowing-boats with a dipping lugsail. This rig required particular skill to handle as the yard had to be passed, 'or dipped', around the mast when staying. The high foot to the sail (possibly reefed) would facilitate the carrying of lobster-pots. Although the dipping lug was superseded to a large extent by the gaff rig, it survived as the traditional rig of many Long Island boats down to the 1950s, and was referred to locally as the 'fore-and-aft lug'. This boat, called the *Maryboat*, was a 'longboat' owned by Mike 'Seán Dónaill' O'Regan of Long Island. She was photographed in 1949.

stem-post. No bowsprit shrouds were used. Two different types of topsail seem to have been in use. One was a four-cornered sail set on a balanced yard. The yard was peaked at a high angle, but its heel extended a short distance forward of the mast. This was referred to locally as a 'lug topsail'. The other topsail, which was used on some boats in later years at least, was triangular in shape and was set on a standing yard which was held in a vertical, or near vertical, position by the well-hardened tackrope.

This is how one old lobsterman described the *Safe Return*, a lobsterboat from East Skeam Island which was reputed to be a particularly good sailer: 'Jim Barry's boat was a beautiful thing to see. She would sail like a yacht. Jim kept her in the best possible shape. Her sails, lines, everything, were the best. There were no knots; every line was a neat splice. In light wind when we would be rowing, Jim would come along with three sails and sometimes a topsail on her and go right on past us . . . she had a beautiful shape from bow to stern. She sat on the water like a seagull.'[16]

As already mentioned, many of the Long Islanders carried a completely different rig. Traditionally, the fishermen of Long Island depended mainly on what were basically rowing-boats with a dipping lugsail. Most of them were four-oar boats about twenty-two feet in length. A few, known as 'longboats', were six-oar boats. The dipping lug was of a much more basic design than the gaff rig, but it required particular skill to handle. It consisted of a single, large lugsail which, unlike the standing lug, extended well ahead

This photograph illustrates the rowing positions in the boat. The two oarsmen would normally be seated, the flood oarsman on a removable seat and the bow oarsman on the port gunnel. The flood oar was known by its Irish term '*béal tuille*'. The helmsman would also row, if necessary, by using the 'hind' oar on the port side, i.e. on the same side as the flood oar. On this boat, the *Valetta*, are Denis O'Donovan on the bow, his brother and owner of the boat, John, on the *béal tuille*, and their uncle Denny O'Donovan on the helm, all from Heir Island.

of the mast, right to the bow of the boat. This meant that the sail and its long yard had to be passed, or 'dipped', around the mast when staying. No shrouds were used; instead, the halyard was taken out to the weather rail, if necessary, to counteract the pressure on the mast. In the bow, the tackrope was often set up with a tackle to get a taut luff. Locally, this rig was known as the 'fore-and-aft lug'. Although, from 1900 on, many Long Islanders had gaff-rigged boats built which were very similar to the Heir Island boats, the dipping lug survived on some Long Island boats right down to the early 1950s.

As well as the two thaufts in the lobsterboats, there were also two removable seats, one in the fore-sheets and one in the stern-sheets. When rowing, the one in the fore-sheets was occupied by the flood oarsman who sat on the starboard side of the boat and used his oar over the port side. The flood oar was referred to by its Irish term '*béal tuille*' (hence the terms '*béal tuille* sheets' and '*béal tuille* thauft'). The bow oarsman, positioned forward of the flood oarsman, used his oar over the starboard side. In most boats, the bow oarsman sat on the port gunnel, but the Long Island boats did not have the usual short foredeck and so had room for an extra removable seat. The removable seat in the stern-sheets, known as the 'seat-board', was occupied by the skipper who steered while his crew manned the oars. The seat in the fore-sheets was known in Irish as the '*clóthairt*'.[17] In some boats at least, the *béal tuille* oar, being used in a wider part of the boat, was longer than the bow oar. In the *Mary Kate*, which had a registered overall length of twenty-seven feet and a registered breadth of seven feet, it is said that the *béal tuille* oar was eighteen feet in length. In general, however, accounts of the length of the oars vary from sixteen feet to twenty-one feet. A spare oar was also carried, and the helmsman could, if necessary, row with this. Facing forward, he pushed the oar with one hand and kept his other hand on the tiller. This oar was usually shorter than the other two and was referred to as the 'hind' oar. It was used over the starboard side of the boat to help the flood oarsman counteract the bow oar which, being positioned further ahead, would have exerted more sideways thrust on the bow of the boat than the flood oar. Row-locks took the form of a pair of thole pins.

The boats were generally coated with Stockholm Tar from the waterline down, inside and outside. Their topsides were usually painted, with the sheer strake given a different colour to that of the hull. On the inside of some boats, the holt was tarred right up to the gunnel, but the fore-sheets and stern-sheets were painted from the floor up. Many of the Long Island boats were completely tarred, inside and out. The gaff and mast were treated with linseed oil except for the masthead which, from the hounds up, was painted the traditional white. The sails were made by the fishermen themselves from heavy calico or cotton duck. Certain fishermen had a particularly good reputation as sail-makers. The calico was white in colour but after one or two seasons fishing, the sails were barked to preserve them. The bark[18] was boiled in water and the resulting solution was brushed on with a soft brush. This gave the sails their usual tan colour. Anchors and all iron fittings were made by local blacksmiths. Itinerant tinsmiths, who were frequent visitors to the islands as well as to mainland homes, played their own small, but important, part in fitting out the lobsterboats. They manufactured tin bailers specially designed to fit deep down into the narrow bilge. Shaped in the fashion of an inverted, truncated cone, they had a narrow, circular bottom and a wide, circular mouth and were fitted with a handle on the side. They had a volume of about one gallon.

Many of the lobstermen themselves were fitted out with oilskin aprons or long smocks made from calico and, in later years, from flour bags. They were waterproofed by applying several coats of a mixture of linseed oil and egg yolk. Between coats, they were rolled up and put aside to allow the oil to soak into them before being put out to dry. It was the egg yolk that gave them their traditional yellowish orange colour, still a feature of modern-day 'oilskins'. Prolonged exposure to sunlight had a detrimental effect on them, gradually eroding their water-resistant qualities. Beeswax was also sometimes used for waterproofing. Some lobstermen also wore the leather, knee-high boots more commonly worn by the mackerel fishermen. They had a strap at the side, near the top, for pulling them on, and they were made specially for the fishermen by local shoemakers. They were regularly treated with oil or goose grease.

4 · The Fishing and the Fishermen

The boats of Long Island, West Calf and Middle Calf generally fished from Cape Clear or the Fastnet 'up north' as far as Dursey Head, as well as locally in the bay. They usually made one-week trips, returning home at weekends, although some made trips of two weeks' duration. The boats from Heir Island, East Skeam, West Skeam and Aughadown usually preferred to fish 'down east' if the wind was favourable, but in actual fact they covered the entire coast from Dursey Head to Ardmore, Co. Waterford. Dursey Head lies thirty-two nautical miles[19] to the north-west of Heir Island and Ardmore over seventy-three nautical miles to the north-east. This 105-nautical-mile stretch of coast represents an incredible mileage of actual shoreline when one considers that the boats fished practically every cove, inlet and offshore reef, not to mention the long shores of Dunmanus Bay and Bantry Bay. These boats would be away from home for periods of up to six or seven weeks, and occasionally even longer.

In late April each year, the boats, having been laid up for the winter, were painted, tarred and launched. Having dried out considerably while laid up, the planks of the boats would have shrunk. On being launched, they were allowed to sink and soak until such time as the planks had swelled again and all leaks had re-sealed. The boats did not sink to the bottom as the buoyancy of the timber kept them 'floating between two waters', i.e. the tops of the gunnels just level with the surface of the water. After several days soaking, they were bailed out, ballasted and rigged. The lobstermen then set out for Baltimore to buy bait, in the form of mackerel, from the mackerel boats landing there. This quest for bait was known as 'scoffing' or 'scorring', words presumably derived from the Irish word *sgaran*,[20] which was the local term for damaged or 'broken' mackerel traditionally given free of charge to the lobstermen. The bait was salted in barrels for the season, with great care being taken with it as it was considered that the best bait for lobsters was that which was also fit for human consumption.

From early in May, weather permitting, the boats would set off in groups of two or three with a crew of three in each boat. The crew of any particular boat would never include more than two members of the one family in order to avoid the risk of a family being left without a bread-winner if misfortune was to befall any one boat. Before setting out, the crew, the lobsterpots, the boat and the sea around it were blessed with Holy Water, and a bottle of Holy Water was also stowed carefully near the stem of the boat to protect the crew from the dangers that lay ahead.

Each boat fished from twenty-four to thirty lobsterpots, depending on the size of the boat. The pots were made of willow withies, and were joined together in a long line by ropes of equal length to form what is referred to as a 'train of pots'. The rope between every two pots, usually fourteen fathoms in length, was referred to in Irish as a *badhilsgith*.[21] This term went out of general use as English replaced Irish as the spoken language of most of the islanders in the final decades of the nineteenth century. It survived, however, among the Irish-speakers of Cape Clear. Three pieces of bait, consisting of half a salted mackerel each, were tied around the mouth of each pot on the inside. The train of pots was stowed in the boat in such a way that on the fishing grounds the pots could be 'shot' or dropped overboard in the correct order as the boat moved ahead, leaving a long line of pots stretched out on a rocky seabed. Each pot was ballasted with two stones tied to its floor on the inside. When shooting close to the shore, the boat was rowed. In open waters, the pots could be shot under sail if the wind was suitable. They were recovered by means of a buoy and buoy-line at both ends of the train, allowing it to be hauled from either end. The buoys were made up from a number of cork net-floats lashed together.

The fishing routine involved hauling the pots about every two hours throughout the day and right through the night. The periods of time around dawn and just before nightfall were known as 'the morning change' and 'the evening change', and the lobstermen always hauled at these times. These hauls were known as 'the change hauls', and they were by far the most productive. Combined with the one or two other hauls made during the short summer nights, they produced over eighty per cent of the daily catch.

When hauling, the two crewmen would row the boat with the bow and flood oars. The skipper stood aft of the main thauft. Having picked up the buoy at the end of the train of pots, he would haul the pots into the boat as she was being rowed along. As nobody was now steering the boat, it was up to the oarsmen, with the help of instructions from the skipper, to keep the boat heading along the same course as the train of pots, which was not an easy task in strong wind. Fishing pots in an area of strong tides, such as the Old Head of Kinsale for example, was particularly hard work for the man who was hauling. Therefore, the slack tide at low and high water was always taken advantage of, especially in deep water where more than one pot would be off the seabed at any one time while hauling. The skipper removed any lobsters, crayfish or crabs from the pots as he hauled them on board one by one, and re-baited them if necessary. Some lobstermen had a custom of spitting on the first lobster taken from the pots and praying, 'May the Lord send us more'. Great care was taken to ensure that the pots were always well baited. After some hours in the water, the bait tended to get soft and mushy. Any such baits were always replaced, especially at the evening change haul in preparation for the heavy fishing usually experienced during the night. The bait was also eaten by crabs, for which there was no market, and they were considered a nuisance. The male crabs were destroyed rather than returned alive, but most fishermen considered it bad luck to kill a female crab because its legs were said to resemble Our Lord's fingers. As the lobsters and crayfish were removed from the pots, they were placed in the stern-sheets and covered with a wet piece of sacking to keep them cool and moist. They were later packed in baskets after their claws were cut at the joints to prevent them killing each other. Many of the Long Island boats had jute bags for storage rather than baskets. The pots were stowed in the order in which they were hauled to avoid any foul-up when shooting. Lobsterpots, like all fishing gear, were always hauled and shot over the starboard side. This was because all true fishermen knew that when Our Lord ordered Simon Peter to cast his nets for the miraculous catch of fish, it was over the starboard side of the boat he cast them.

Once the pots were hauled, they were moved to another rocky patch of seabed along the shore or, during daylight hours, out to deep water where they were shot on a submerged reef or rock, which was referred to by the Irish terms *rabha* or *builg*. While the pots were fishing on the bottom, the boat was anchored in the nearest sheltered area. A 'pole' was the form of anchor most commonly used. This was an oblong stone about twenty-four inches in length and scored in the centre to take the rope. It was particularly useful when anchoring in a rocky area where one could easily lose an expensive iron anchor.

Great care was taken to keep the catch alive until it could be sold at the end of the week. While the boat was at anchor, the baskets in which the lobsters and crayfish were stored were suspended in the water 'to give them a drink'. Each boat carried two baskets measuring approximately two feet square and three feet in depth. Like the lobsterpots, they were made from willow withies. The mouths of the baskets were covered with netting. A length of rope, known as the 'ray', stretched across the top of each basket and was interwoven half-way down both sides to guarantee a secure hold. The basket was suspended by another length of rope tied to the centre of the ray and leading in over the gunnel, and made fast to the reesing[22] of the boat. The average catch, per boat per week, was in the region of twenty-three or twenty-four dozen fish, most of which would have been lobster rather than crayfish. This was an enormous catch by today's standards. The baskets could hold from eight to eleven dozen each. When they were full, the remainder was held in jute bags which would hold about two dozen each. In the event of an unusually large catch, some pots were removed from the train and used for storage.

One of these baskets full of lobsters and crayfish was a considerable weight, and a certain technique was required to bring it on board when it was time to haul the pots. Nearly half-way down either side of the basket was a strap or handle made of rope. Two of the crew caught a handle each. Tilting the top of the basket away from them, they brought the bottom of it up against the side of the boat and held it in that position. The third man then heaved on the rope leading from the ray. With the bottom of the basket pivoting against the side of the boat, the top of it swung up and out of the water and the basket landed gently on its side across the gunnel of the boat. It was then carefully lifted down into the boat.

Bantry in the 1890s, from the Lawrence Collection. In the foreground is a train of lobsterpots. In the midst of the pots is one of the baskets used for storing lobsters and there is another one on the left.

5 · Life Under the Towel

The only shelter available to the lobstermen in their entire six or seven week fishing trip was a tent of sorts, temporarily erected in the bow when the boat was at anchor. The tent was originally referred to in the local Irish dialect by the term *teabhal*.[23] Over time, this word was confused with the English word 'towel', which is similar in pronunciation but does not seem to correlate in any other way with *teabhal*.

The towel was made of calico or a smooth canvas known as 'baft'. It was draped across a boat-hook or sometimes an oar, one end of which rested on the stem of the boat or on the short foredeck, and the other end of which was tied to the mast at a height of five or six feet. Each towel was made specifically to fit the boat in which it was used. It extended out over both gunnels, and was wrapped around a plank of wood at either side to weight it down on the outside of the boat. The end of the towel was brought back up, between the plank and the sheer strake, over the gunnel and into the boat. The weight of the planks was usually enough to keep the towel in place, as the boat would be lying to the weather when anchored, with the open mouth of the tent facing away from the wind. In a gale, however, it was also lashed down inside to the reesing. The planks of wood were known as 'thaufts', the same term that was given to the thwarts of the boat. On some boats, the towel was reinforced with a length of extra heavy canvas where it rested on the oar or boat-hook. This piece of heavy canvas was known as a *srathair*,[24] the Irish for straddle.

In the nineteenth century, the towel was a common sight on all open fishing boats on the coast of West Cork,[25] but it survived in the lobsterboats long after it had gone out of use on other boats. Besides the towel described above, there was also a small towel which extended only half-way from the stem to the mast. The lobsterboats, however, always carried a big towel, and in the lobsterboats it developed a second function. Triangular in shape and larger than a staysail, it was reinforced by a rope sewn on all around its edge with a cringle at each corner so that it could be jury rigged in the fashion of a spinnaker when running before the wind. This use of the towel as a sail was a particularly common practice in the Heir Island boats. Hence, the local term for these boats: 'the towelsail yawls'.[26]

Cooking was done at the mouth of the towel, on an open fire contained in a bastible pot,[27] just forward of the mast and the main thauft. The pot, referred to as the 'fire-pot', was set on a mound of white clay under which was a large flat stone or a sod of earth. The bottom of the pot was also lined on the inside with a further layer of white clay or gravel. On this layer of clay or gravel, a fire of coal and driftwood was built. In some boats at least, the fire was set alight with the aid of a hand-bellows, and a piece of corrugated iron, standing abaft the fire-pot, protected the thauft and the mast. Bread, baked in a second bastible set on the fire-pot, was the mainstay of a lobsterman's diet. Two cakes were baked for each of the three daily meals, spread with butter and eaten hot. So-called 'griddle cakes' were baked in a frying pan. The speciality of Dan Harte, skipper of the *Hanorah*, was crab cake. He frequently added the red meat from a crab to his cake mixture. Handlines and crab bait or mackerel bait were used to catch pollock, bream and wrasse, which is known locally as 'conor'. White conger eels, which were frequently caught in the lobsterpots, were preferred salted. Black congers, considered inferior in quality, were seldom eaten. The

The only shelter on the lobsterboats was a tent, known as a 'towel', erected in the bow when the boat was at anchor. This towel could also be jury rigged in the fashion of a spinnaker, hence the term for these boats, 'towelsail yawls'. The lobsters and crayfish were stored in two baskets suspended from the boat. The ropes holding the baskets can be seen in this photograph (from the Fr Browne Collection), one near the shroud and the other further aft. In this photograph, taken in Cork Harbour, a local rowing boat has pulled up to the far side of the lobsterboat.

salted mackerel used for baiting the pots was sometimes eaten, preferably after it had been in the pots overnight so that much of the salt had dissolved. It was usually resorted to only in times of food shortage, as salted mackerel was considered the best lobster bait, and lobsters took priority over crew. The fish was usually boiled, and if potatoes were available, they would be boiled in the pot together with the fish. Meat was rarely eaten. In the days before it became fashionable to have a floor in the holt, the pot of boiled potatoes was tipped on to the ballast, and the crew ate directly from the mound of steaming potatoes without the use of plates. In the nineteenth century, yellow-meal porridge, or *praiseach*, was also frequently eaten, especially when potatoes were scarce.

In later years at least, tea, sugar and bread soda were among the food items stored in the locker in the bows, along with Sunday clothes. Cans and enamel mugs were used for drinking, and the tea was brewed in a kettle. Milk was contained in a four-gallon earthenware whiskey jar, which was sometimes fitted with a protective sheath woven from straw. Keeping the flour dry was a constant concern. Although the towel was barked to preserve it, waterproofing it was considered a waste of oil as it was frequently exposed to strong sunlight which would quickly erode its water-resistant qualities. However, drops forming on the inside of the towel could be diverted down to either side by running one's finger down along the calico. When cooking was in progress, the bag of flour was put under the main thauft near the fire. When the crew was sleeping, it was covered over and used as a pillow. When the towel was not in use, the bag of flour was wrapped up in it together with the straw or heath on which the crew slept. Water was stored in a wooden cask of about ten gallons, known as a 'breaker', which had a rope tied around it for carrying it. It was laid on its side, and had a bung in its centre under which was tacked a leather spout.

When lowered, the mainsail was stowed by peaking the gaff up fully and standing it vertically against the mast. The sail was then pleated or furled against the gaff, and the peak halyard was wrapped around the sail, gaff and mast to keep the entire arrangement in place. The oars, being so long, were difficult to stow when the towel was in use. Usually, the two longer oars were stowed by slipping one end of them inside a loop of the peak halyard fairly high on the mast, where it was wrapped around the furled mainsail. One could then easily walk under this end of the oars, the other end rested on the seat-board in the stern-sheets. The hind oar was stowed to one side of the boat with the end of it extending out over the transom. Space also had to be found for a bag of coal and a small barrel of lobster bait. Alternatively, the bait could be wrapped in canvas and stowed under the seat-board in the stern-sheets.

With all these supplies and accessories, together with anchors, warps and other paraphernalia (not to mention the train of pots

Cooking and baking was done in a bastible pot (visible in this photograph) set on an open fire contained in a second bastible, known as the 'fire-pot'. Straw, or sometimes heath or hay, was used for a bed. This photograph, also from the Fr Browne Collection, was taken in Cork Harbour in 1928.

and baskets of lobsters that would be on board when making a passage), the stowing of equipment had to be well planned. This was reflected in one man's directions to the boat builder who, when ordering a new boat, stipulated that she was to be twenty-seven and a half feet in length and seven feet in beam with the height of a jug from the floor to the main thauft and the height of a pint mug from the thauft to the gunnel. The jug referred to was the earthenware jar, or *crusca*, for milk which was usually stowed under the main thauft.

Ciarán Ó Síothcháin's drawing (IFC 1320, p. 457) of a lobsterboat showing its different parts and the ridge-pole for the towel (usually a boat-hook or an oar) in position.

1. *Corp an bháid:* The hull of the boat
2. *An crann:* The mast
3. *Stá an chrainn:* The mast stay (forestay)
4. *An striúda deas:* The starboard shroud
5. *An striúda clé:* The port shroud
6. *Tochta bhéal tuille, no tochta crainn bhéal tuille:*
   Flood thwart, or mast flood thwart (usually referred to as the *béal tuille* thuaft or the main thauft)
7. *An tochta deirig:* The aft thwart (or hind thauft)
8. *Caipín an stuimne:* The stem cap (or stemhead)
9. *An siota deirig:* The stern-sheets (or hind-sheets)
10. *Siota bhéal tuille:* *Béal tuille* sheets (or fore-sheets)
11. *Maide droma an teabhail:* The ridge-pole of the towel

The drawing does not show the locker in the bow which was a feature of the Heir Island boats. Most boats had two shrouds at either side.

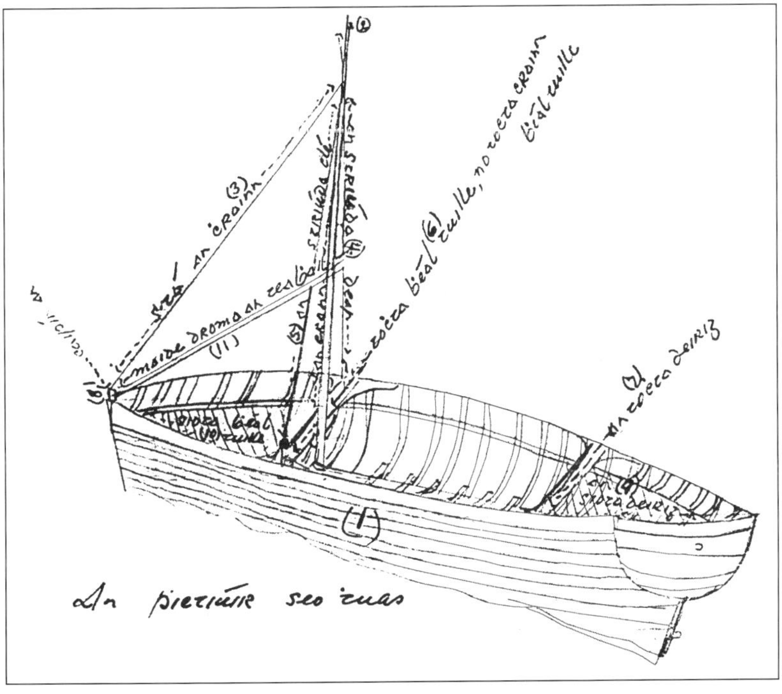

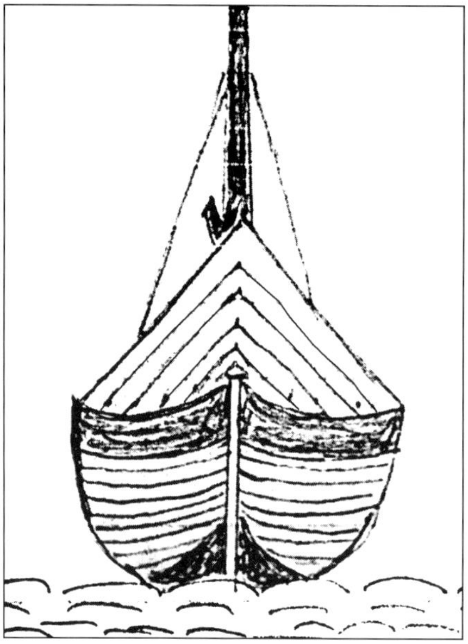

Ó Síothcháin's drawings (IFC 1320, p. 492 and p. 493) showing a towel without a straddle (*left*), and with a straddle (*below*) viewed from ahead.

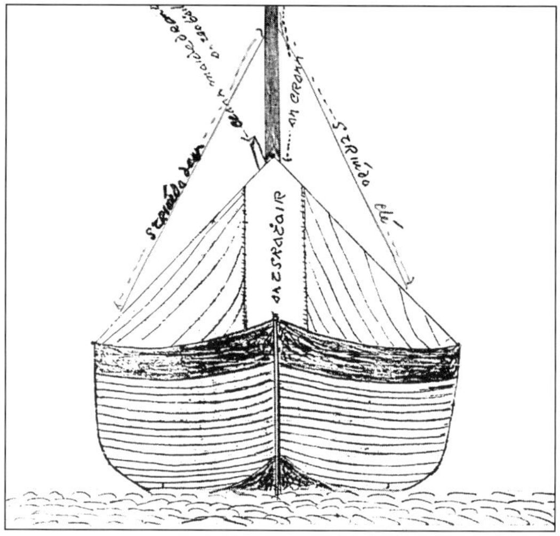

(*below*) Ó Síothcháin's drawing (IFC 1320, p. 495) of a lobsterboat at anchor with the towel erected.

1. *Téad phóill*:	Pole rope (the pole was an oblong stone used as an anchor. It was scored to take the anchor rope or pole rope)
2. *Srathair*:	Straddle (a heavy length of canvas that reinforced the towel where it rested on the ridge-pole)
3. *An teabhal*:	The towel
4. *Béal an teabhail – An dul isteach fé*:	The mouth of the towel – To go under it
5. *Ceann maide dromadh an teabhail*:	The end of the towel ridge-pole
6. *Píce an teabhail*:	The peak of the towel
7. *Barra an chrainn*:	The masthead

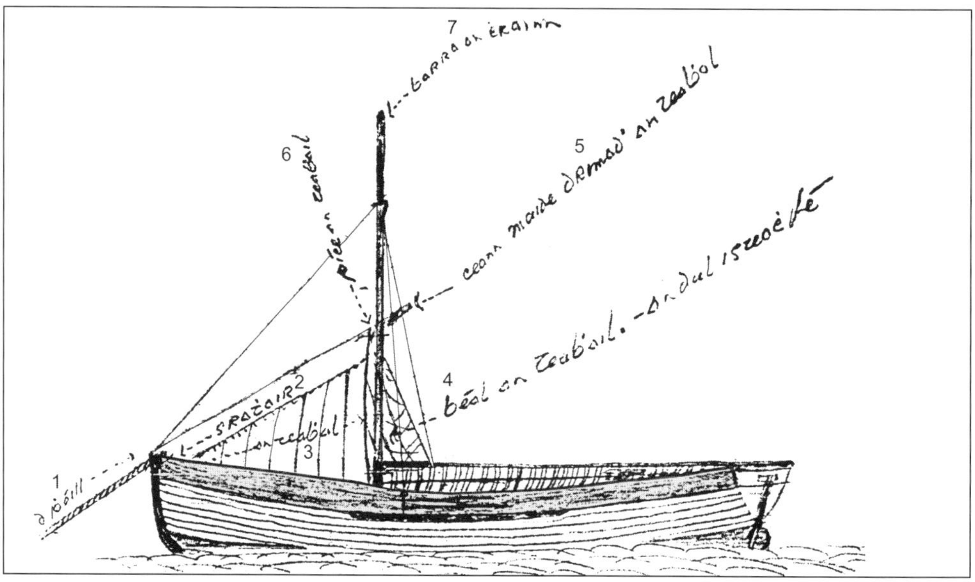

6 · Inherited Wisdom

Lacking chart and compass, the basic requirements of conventional navigation, the lobsterman depended on an extensive mental reservoir of inherited wisdom that enabled him to fish his way along the coast. Submerged reefs were located by the age-old method of aligning sets of known landmarks to position the boat over the reef. It is also said that the older, more experienced lobstermen could find these reefs by instinct in thick fog when the 'marks' were not visible. These marks, together with an intimate knowledge of the shoreline itself, including tideways, wrecks and navigational hazards, were part of a vast store of knowledge accumulated and passed down by successive generations of lobstermen.

The experience of generations had also given the lobsterman that understanding of the habits and movements of his quarry that every successful hunter needs to have. The early stages of a flood-tide, for example, is the best time to catch lobsters close to the shore in a *cuas*.[28] The favourite haunt of the crayfish is the high ledges of the offshore reefs, where they can be caught at 'the slack of the tide'.

The lobsterboats carried no navigation lights, fishing lights or deck lights of any description. The lobstermen, therefore, had to be able to remove lobsters, crayfish and crabs from the pots in total darkness, not to mention the far more dangerous conger eels. The towel was illuminated, if necessary, by an oil lamp or by a candle stuck to the reesing of the boat with candle grease. In the nineteenth century, candles were often homemade from tallow which was salvaged from shipwrecks or washed up along the shore. The tallow was melted down and poured into a tin cylindrical mould around a wick of white cord. The mould had a screw cap at one end which was removed when the candle had set so that it could be pushed out.[29] These moulds were manufactured by the same itinerant tinsmiths who made the bailers for the boats. A basic oil lamp, consisting of a seashell containing fish oil, was sometimes used. A dried rush with its skin peeled off acted as a wick. Eventually, the seashell was replaced by a tin container, also manufactured by the tinsmiths, but it continued to be referred to as a *slige*, the Irish for shell. Long narrow strips of bog fir were also used as tapers. They were hand-held and had to be inverted from time to time to keep them alight.

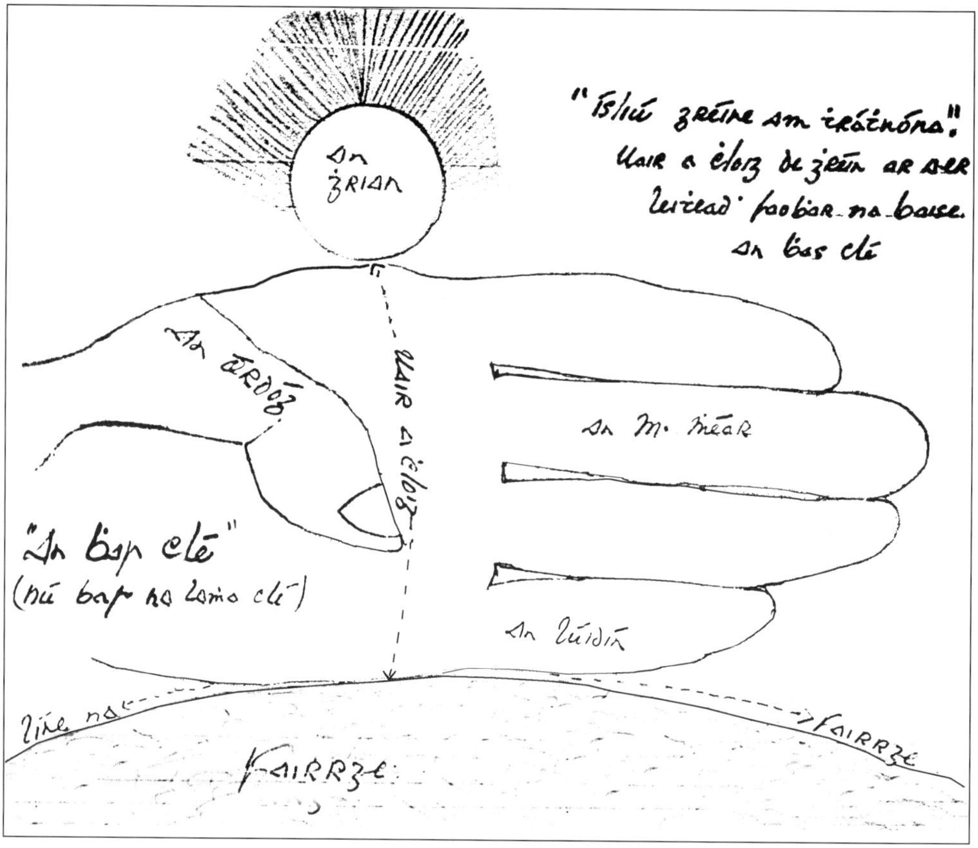

Ciarán Ó Síothcháin's diagram (IFC 1357, p. 285) depicting how the height of the sun was gauged to calculate the time remaining before sunset. The width of four fingers between the sun and the horizon meant one hour until sunset. Each finger accounted for fifteen minutes so that, using both hands, the last two hours before sunset could be measured in fifteen-minute intervals.

Timepieces of any description rarely figured among the possessions of ordinary people in the nineteenth century, and it was well into the twentieth century before watches were carried in small fishing boats. Instead, fishermen measured time by gauging the height of the sun in the sky. The width of four fingers (with one's arm at full stretch and not including the thumb) between the horizon and the evening sun meant that the sun was within one hour of setting. If the four fingers of both hands were required to fill the space, then it was two hours to sunset. Each finger accounted for a quarter of an hour so that the time remaining to sunset could be gauged to within fifteen minutes.

Very often, the lobsterman's life depended on his uncanny ability to forecast the weather by careful and constant observation of the world around him. The sky, the moon and stars, the sun, the tides and swell, the phosphorescence in the water at night, the behaviour of seabirds and, indeed, the behaviour of the lobsters themselves, were all natural barometers to those who knew how to read them. Lobsters, for example, become very agitated and aggressive when a storm is approaching, snapping and biting when being removed from the pots. A shag flies high only when bad weather is on the way. Not only that, but it always flies in the direction the weather is about to come from. The presence of *lóch*[30] in the pots as they were being hauled or of fleas in freshly plucked heath for bedding were sure indicators of impending gales.

One can imagine that in such an environment, where a man depended for his very survival on observing what was going on around him, that his perspective would extend beyond the realms of the natural and into the world of the supernatural. Just as he interpreted natural phenomena, or 'the signs', to predict the weather, so also were there 'signs' by which he predicted his fortune. Just as bad weather followed a rising swell, bad luck followed the sight of a hare, a fox or a red-haired woman on the shore. Dogs or even otters, which are known locally as 'water dogs', also brought bad luck to some. A boat's good luck could inadvertently be given away simply by one of its crew giving an object to a crew member of another boat. The object would have to be recovered, and with it the boat's luck, before fishing could continue. The giving of 'a light' on Monday was particularly bad luck. Side by side with these beliefs on the part of the lobstermen was a steadfast faith in the protective powers of the Holy Water carried near the stem of the boat; and every evening, while at anchor, the men would kneel to pray as the skipper led them in reciting the Rosary.

7 · Odyssey

The lobstermen, having set off from home around the middle of May, would fish their way along the coast or, if possible, they might take advantage of a fair wind to head directly for one of their favourite fishing grounds. A boat leaving Heir Island on a Monday morning might set sail for the Old Head of Kinsale, for example, and have the first shot of the season that evening at the Cuainín, west of the Old Head, over thirty-five sea miles from home. There was a strong belief that once a boat set out to make a passage, she should not return to her port of departure without reaching her intended destination, especially if she set sail on a Monday.

Having started the season at the Cuainín, a crew might spend the first week fishing their way east around the Old Head and north to the mouth of Kinsale Harbour. Early on Sunday morning, they would put into Kinsale to attend to their religious duties, and also to be ready to sell their week's catch early on Monday morning.

Lobsterboats in at Kinsale to sell their lobsters. Kinsale and Cobh were the two ports at which the lobsters were sold. This photograph, from the Greene Collection, was taken sometime between 1910 and 1925. The lobsterboat furthest from the pier appears to be Con Minihane's *Leader*.

Kinsale and, from at least as early as 1920, Queenstown (later renamed Cobh) were the two ports where the lobsters were sold. Every Sunday morning would see up to seventeen or eighteen of the boats tied up at the quays in each of these harbours. The lobsters were sold by the dozen, rather than by weight, but any small lobsters in the catch were counted separately, and every two of them considered the equivalent of one of acceptable size. In the 1880s, they were making ten to twelve shillings per dozen. During World War I, the price is said to have fallen as low as four to six shillings. In the 1920s and 1930s, lobsters were back up to ten shillings per dozen and crayfish were twelve shillings.

The lobstermen would avail of some recreation in the pubs during their short stay in port. Many of the Heir Islanders had the reputation of being excellent singers, and they could frequently be heard singing in the pubs or on their boats tied up alongside each other

A photograph from the Fr Browne Collection showing two towelsail yawls at Cobh in 1934. The top of one of the submerged baskets for storing the lobsters can be seen suspended from one of the boats. Among the items faintly visible on board the outside boat is the earthenware jar, or jug, for milk which was stowed under the main thauft (thwart). It is near the coil of rope hanging from an anchor on the starboard side of the boat. The *Heber* (left), owned by Mike O'Driscoll of Heir Island, was one of the mackerel yawls that occasionally joined the fleet to fish for lobsters when the market for mackerel was poor. O'Driscoll's storage baskets are suspended from buoys, made up from net-floats.

Two towelsail yawls at the Camber in Cobh, thought to be around 1938.

at the pier. Meanwhile, their womenfolk back home on Heir Island would be looking forward to the first in a series of letters containing some of their well-earned wages. The arrival of the first such letter of the season would usually mean a trip to Ballydehob pig fair for the woman of the house. Rowing or sailing the four miles to Ballydehob in their small lugsail yawls, the women would each buy two bonhams, one to be fattened for sale and the other to be slaughtered at Christmas and salted for home consumption.

Once their catch was sold on Monday morning, the lobstermen would put to sea once more and the following week might be spent fishing between Kinsale and Cork Harbour. Further stages of their odyssey might take them right on east beyond Roches Point and Power Head to Ballycotton, Youghal and Ardmore, Co. Waterford, but they always returned to Queenstown or Kinsale to sell their lobsters.

Fresh water, needless to say, was one of the most valued commodities on board, and it is said that there was not a well or a spring along the coast from Ardmore to Dursey Head unknown to the lobstermen. A well at Power Head, east of the mouth of Cork

Towelsail yawls dried out at the Camber in Cobh. The locker in the bow, in which food and clothing were stored, can be seen in the boat in the centre.

Harbour, was a useful source of water when the wind was from the east. It was reached by means of steps on the lee side of the headland, leading up from sea level to a fog signal station. At the east side of the Old Head of Kinsale, steps cut into the rock in the cliff face, in the *cuas* directly inside of Bream Rock, gave access to another well. In addition, paraffin oil could be had from the lightkeepers to start the fire. A little cove near the mouth of Oyster Haven, known as Walker's Cove, had a well on the shore fed by a spring higher up in the cliff. The Walker family, who lived nearby, was among the many farmers along the coast who traditionally supplied milk and potatoes to the lobstermen. One member of this family actually went as far as providing for the lobstermen in his will, stipulating that his beneficiary should always make milk available to them. The Howe farm near Howe's Strand in Courtmacsherry Bay and, in later years, White's farm near the mouth of Kinsale Harbour, were also generous sources of supply. Fresh straw for bedding was also available from farmers. Heath, plucked from hillsides along the shore, was also used.

Another cove at the west side of Oyster Haven was a favourite anchorage in bad weather, and the sight of lobsterboats anchored inside of Woodpoint near Courtmacsherry was taken by the locals as an indication that they should secure their boats against imminent strong winds. When sheltering in Ringaskiddy, in Cork Harbour, some boats would shoot their pots right in at the old pier, mooring the end of the train of pots to the ladders on the pier as a precaution against the strong tide. Left overnight, they always caught some lobsters before morning. The boats also fished up into East Ferry where there was a pub frequented by the lobstermen. During World War I, the *Flower of May*, from the West Skeam Island, was forced into Cork Harbour one night by bad weather. Entering the harbour after curfew, she was commandeered by the Navy, and her skipper Rickie O'Regan and his crew were detained at Haulbowline Navy Base.

Occasionally, boats were weather-bound for long periods. Sometimes, finding themselves trapped on an exposed stretch of coastline in the face of a strong gale or storm, the crew would be forced to remove all the equipment and ballast and, with the aid of locals, haul their boat up on the shore until the weather had improved. Around the turn of the century, the crews of three Heir Island boats were given accommodation in the convent in Ardmore, Co. Waterford, having had to haul up their boats. Dan O'Regan of the West Calf once had to haul up his boat, the *Ellen*, on Dursey Island.

8 · Trials and Tribulations

Early on Sunday morning 5 August 1945, two Heir Island lobster-boats, the *St Margaret* and the *Mary Kate*, were anchored near each other off Broadstrand on the west shore of Courtmacsherry Bay. They had spent the week fishing the several reefs in the bay, including the Big Barrels, the Small Barrels, the Bluebell, the Black Tom and the waters around Horse Rock. To the inexperienced, it was a fine morning to make the thirteen-mile passage east across the bay, around the Old Head and north into Kinsale Harbour. To the lobstermen, all the signs predicted strong winds from the north.

Anxious to make Sunday Mass and to sell his lobsters, Jim McCarthy, skipper of the *Mary Kate*, was of the opinion that he would make Kinsale before the weather broke. Leaving John Harte and the *St Margaret* anchored at Broadstrand, he set out in a light northerly breeze to haul his pots which were shot at the Barrels in the middle of the bay. With his pots on board and heading for the Old Head, his crew of Timmy Murphy and young Jack Pyburn reefed the sails in the strengthening wind. Half-way to the Old Head, they were forced to reef her down another reef. As they anxiously approached the Old Head, an area notorious for baffled winds blowing down off the cliff, they took in their third and final reef. Just as they cleared the headland, they were blasted by strong gale-force squalls ripping down along the east side of the peninsula. One particularly violent, sudden gust broke their gaff in two. As they struggled to lower their mainsail and get their boat under control once more, they knew their lives were in danger. With the now useless mainsail and broken gaff out of the way, they took up their oars for what they knew was going to be a long, hard, desperate struggle. In danger of being blown out to sea, they rowed with every ounce of strength in their bodies, battling the steep waves for every foot of headway. Eventually, nearing exhaustion, they reached the shelter and safety of the *cuas* inside of Bream Rock on the east side of the Old Head. They rowed right into the end of the *cuas*, anchored, rested and waited.

At nightfall, the wind died down and the sea calmed. Jim McCarthy lashed the broken gaff to an oar and they set out once more for Kinsale. Rowing, in the now almost dead-calm conditions, they reached the Bulman Buoy near the mouth of Kinsale Harbour late at night. They shot their pots in a *cuas* inside of the Bulman and anchored. When they hauled again at about 4 a.m. on Monday morning, it was still dead calm. After shooting their pots once more, they rowed the remaining two miles into Kinsale against an ebb-tide.

Arriving in Kinsale Harbour on the morning of regatta day, the highlight of the year in the social calendar of people for miles around, they knew that their chances of finding a carpenter willing to set to work on their broken gaff were slim. They sold their lobsters, but it was late the following evening before their gaff was repaired. It was Wednesday morning before they were ready to set sail once more.

Such were the trials and tribulations that sometimes befell the lobsterman.

Towelsail yawls at Union Hall. The lobstermen had an uncanny ability to forecast the weather and usually made the shelter of some harbour before the arrival of strong winds. In the first photograph, two lobsterboats are moored side by side with their stems to the pier, towels erected and their pots on board. In the second photograph is a total of seven lobsterboats. The pots appear to have been put ashore on the pier.

9 · First Trippers

Most young Heir Island boys looked forward eagerly to the day they would set off on their first trip in the lobsterboats. Mesmerised by adventurous stories of distant harbours, dangerous passages and big catches, they longed to see for themselves the sights of Cork Harbour and marvel at the size of the farms to the east, some of which were larger than their entire island.

In the meantime, these boys learned the skills of lobstering from the older men, the men who had spent a lifetime travelling out in the lobsterboats but who were now no longer able for the hardships of the long trip. These old men and young boys fished locally in Roaringwater Bay in small sixteen to eighteen-foot yawls with dipping or standing lugsails.

For many first trippers, however, dreams of adventure were all too quickly shattered by harsh reality. Monotonous, muscle-building hours of hard rowing in calm weather was one aspect of the job that was guaranteed to dull the sense of adventure of even the most enthusiastic youngster. So it was with young John Harte,[31] as he toiled on the bow oar of the *Hanorah* off Oyster Haven, east of Kinsale. He understood how another young 'first tripper' must have felt in the cautionary tale his skipper Dan Harte recounted. As everyone knew, to stick a knife in the mast of a boat was bound to bring dangerously strong wind. In the old days, according to Dan's story, a 'first tripper', tired of rowing and desperately hoping for wind, lost his patience, grabbed a knife, jumped up on the main thauft and, reaching as high as he could, drove it into the mast with what force he could muster. The resulting storm almost sank the boat. In spite of this warning, John Harte was tempted to experiment with a knife.

Before temptation got the better of him, relief came to him by another means. Pulling hard on his aging oar, he broke it. The tirade of verbal abuse from his crew-mates for his carelessness in breaking the oar could not dampen his joy. Putting in to Oyster Haven, John Harte enjoyed the three days they fished locally as they waited for a carpenter to make a new oar. His joy quickly turned to sorrow, however, when he took up his new oar to resume the long passage. To his dismay, it was made of freshly cut, green timber and his heart fell when he felt its weight. For John Harte, his first trip was going to be a long one.

John Harte was to spend a lifetime lobstering along the south coast. Other 'first trippers' were not as fortunate. The homemade oilskin aprons and long smocks worn by many lobstermen were not as durable or as weather-resistant as modern foul-weather clothing, and the lobstermen frequently worked and slept in wet clothes. At a time when for the ordinary working man (on land or sea) a doctor was only called upon to help the dying, as distinct from those who were merely sick, these harsh living conditions could occasionally prove fatal. The most melancholy case of death on a lobsterboat, however, was that of a young boy on his first trip who suffered a painful death from a ruptured appendix, many miles from his home on Heir Island.

For some young Heir Islanders, a trip down east in a lobsterboat was only the first short leg in a voyage that would carry them thousands of miles from home to even more wondrous places than Cork Harbour. During the War of Independence and the Civil War, rail connections between West Cork and Cork City were constantly disrupted. On one occasion, Mike 'the Cloud' O'Neill of Heir Island cut short his fishing trip down east and returned home to bid farewell to his daughter, possibly for the last time. She and another young girl from the island were emigrating. On returning home, Mike 'the Cloud' learned that the girls had been unable to arrange transport to Queenstown to get the liner to America. After eating a meal, he and his crew re-baited their lobsterpots and set out to the east again with the two girls on board. Sailing through the night, he landed the two girls at the West Camber in Cobh early the following morning, where they stepped out of the world of the towelsail yawl forever and headed for another world in a big American city.

Young Heir Island boys learned the skills of lobstering from the older men who were no longer able for the hardships of the long trip but fished locally in the small, general-purpose, lugsail yawls. This one, about eighteen feet in length, was built at Galley Head in the 1930s. She was bought by Jim McCarthy, Heir Island in the 1950s who sold her to Timothy John McCarthy, also of Heir Island, sometime after. She is now owned by Timothy John's son Dan. In the background of this photograph are some of the houses at Paris, Heir Island, including Pyburn's house (gable on the left). It was here that Richard Pyburn, and his father before him, built the Heir Island lobsterboats and also several smaller lugsail yawls. *Photograph: Kevin O'Farrell*

The small lugsail boats were even more commonly used for lobstering by the Long Islanders. Limited in range, they usually fished locally in the bay, but some of them carried a towel like the larger boats and remained at sea for a week at a time. This boat, owned by Tim O'Sullivan of Long Island, is typical of this type of boat. She is said to have been built by Willie Skinner of Rathmore, Baltimore, but some dispute this strongly. One of the oldest intact work boats on the south coast, she is thought to date from 1882. O'Sullivan bought her from Tommy O'Mahony of Cape Clear in the late 1940s, and fished lobsters with her. It is said that she once brought the corpse and twelve mourners in a funeral from Cape Clear to Baltimore. She is eighteen feet in length. *Photograph: Cormac Levis*

In periods of fine weather, even smaller, one-man boats were fished locally for lobsters, including this one, once owned by Jamsie O'Regan of Long Island. These boats were flat-bottomed and this one, built in 1945, measures 13'8" in length. She was one of several of this type built by Timmy O'Donovan of Schull. In this photograph, she is being hauled up by her present owner, Tim O'Sullivan of Long Island.
Photograph: Cormac Levis

10 · Up North

Early in July, after about six or seven weeks fishing 'down east', the lobsterboats would return briefly to Roaringwater Bay. On Heir Island, 'the grass of two cows' (four or five acres) was the size of a typical holding, although some were up to twelve acres. During their three or four-day visit home, the main task of the Heir Island men was the drawing in of their acre or acre and a half of hay. The hay would have already been mown by farm labourers from the mainland or neighbouring islands, and saved and made into cocks by the women and children while the men were at sea. Once the rick of hay was standing safely in the haggart,[32] they put to sea again and continued lobstering right into October or early November, making trips of up to six or seven weeks' duration to the east, and short trips 'up north'.

A photograph taken in Castletownbere in 1913. The grounds 'up north' were the favourite haunt of boats from West Calf, Middle Calf and Long Island. The fire-pot is just visible ahead of the mast, and an earthenware jam jar used as a cup is also visible. The man nearest the camera is posing with a crayfish in his right hand. *Spillane Collection*

The northern grounds, which were the grounds preferred by the lobstermen from Long Island, West Calf and Middle Calf, extended right out along the shores of the Mizen, Sheep's Head and Beara peninsulas. On one occasion, the *Pride of Toe Head*, owned by Timothy O'Donovan, arrived back home to Heir Island with a phenomenal catch of fifty-four dozen lobsters and crayfish after a one-week trip up north. This was not counting the four or five dozen fish said to have died on the passage home.

The daily fishing routine varied to a certain extent among the boats. The Long Islanders generally seem to have hauled their pots less frequently than the Heir Islanders. Most of them hauled at dawn, at nightfall and once or twice during the day, and very few hauled in the middle of the night. In good weather, some Long Island boats fished right into the sea caves in the Mizen Head, going in as far as the height of the mast would allow. The caves narrowed gradually until a point was reached where a boat could not be turned around, and so they were entered stern first so that the first pot of the train could be shot as far into the cave as possible.

Dunlough Bay, at the very end of the Mizen Peninsula, always offered good fishing but is an extremely dangerous part of the coast in bad weather. At the first signs of the wind strengthening from any westerly point, the boats would retreat to the safer fishing grounds down along Dunmanus Bay, an indication to locals that bad weather was on the way. In strong gales, the boats anchored in the shelter of Blair's Cove, right down at the end of the bay.

On the Beara Peninsula, Blackball Head and Carrigbofinn were among the favourite areas fished. At Clohane Beag, west of Castletownbere, a long narrow gorge leads into a completely sheltered

A young artist's impression of a lobsterboat at anchor in Bank Harbour, on the Beara Peninsula. It was drawn by Richard Power of Bank on 26 July 1906. The towel is up and there seems to be some cooking in progress. The object at the bottom left of the drawing presumably represents one of the baskets for storing the lobsters.

basin where the boats could anchor in safety. No craft much bigger than a lobsterboat could get in there, and it was said that God made it especially as a refuge for the lobstermen.

Cuas an Uisce, east of the Mizen Head on the south side of the peninsula, was, as the name implies[33] one of the many places where fresh water was taken. Water was also available on all of the islands in Roaringwater Bay. On Long Island, there is a folk-memory reaching back into the nineteenth century of several Heir Island boats anchored at Cuasamhuillin while the lobstermen cooked *praiseach* with water from the nearby well. They also took water from another well on Long Island at Cuasach, near the east end of the island.

As well as the grounds up north and down east, the rocks and reefs around the Fastnet, Cape Clear, the Calf Islands, Goat Island and Oileán Móineach, and the Builg Mhór west of it, were also fished. The Long Islanders, in particular, fished these areas, and on a terrible night of thunder and lightening, in the early 1920s, a Long Island lobsterboat lay at anchor near Carrigdwyer, north of the Middle Calf. For some reason, the skipper of the *Maryboat*, Mike 'Seán Dónaill' O'Regan, had only one crewman with him that night, his wife's nephew, Peter O'Driscoll. The two men sheltered under the towel as the thunderstorm raged. They feared for their safety as their boat and the sea around it was brightly illuminated by flash after flash of lightening, with deafening claps of thunder and a loud, lingering crackling as if the very sea was on fire. As it turned out, their fears were well founded for this was to be Peter O'Driscoll's last night at sea.

As he knelt at the mouth of the towel, anxiously observing the terrifying spectacle of the lightening storm, O'Driscoll leaned against the mast. Just at that moment, a bolt of lightening hit the mast and killed him instantly. One cannot even begin to imagine Mike O'Regan's state of mind as he made the harrowing journey back across the bay to Long Island that night. The thunder and lightening raged on as he rowed his six-oar boat single handed through the deluge of rain, bringing his dead relation back home to the families he knew would be anxiously praying for their safe return.

Like all coastal dwellers, the lobstermen were always on the look out for any prize that might be won from the sea by way of wrack. Many boats were built and many houses were roofed with wrack timber. For some, the collection of wrack became an obsession, with people going to extraordinary lengths to wrestle it from the sea. Some, apparently, could even smell it in the air. One night, probably sometime in the 1930s, a Heir Island crew rowed their boat into a *cuas* on the Beara Peninsula. As they rowed, a man by the name of McCarthy, who was on the bow oar, announced that he could smell fresh timber. Making their way carefully further into the *cuas*, they looked out for floating wrack but could make out nothing in the darkness of the secluded cove; and so they shot their pots. As they lay at anchor, McCarthy was still convinced that he could smell timber. In the grey light just before dawn, the skipper, Minihane, picked up the buoy to begin hauling the pots, and was suddenly aware of a large object floating nearby. The object turned out to be a gigantic baulk of wood. After hauling their pots on board, the men tied on to the baulk with a tow rope. All that day and right through the following night, they laboured with their oars, dragging behind them the heavy, awkward bulk of the huge beam. Making very slow headway, they reached Heir Island early the following morning. It is about twenty-five nautical miles from Heir Island to the nearest part of the Beara Peninsula. After selling their lobsters to Con Burke on Heir Island, they had breakfast and set off for Ballydehob quay, four miles away, still towing the baulk. In Ballydehob, they sold the timber to a local businessman, Joe Young, for thirty shillings.

11 · Homeward Bound

From the middle of October to the start of November, the lobster season would gradually come to an end, with the lobsterboats homeward bound for the last time as the lobstermen could no longer expect the weather to provide any prolonged periods of relatively calm sea conditions.

On Heir Island, the most immediate task awaiting the return of the lobstermen was the threshing of the wheat, the main cereal crop grown on the island. The back-breaking work of harvesting the wheat with a reaping-hook and binding it into sheaves would have been carried out by the women in the absence of their menfolk. Although some of the threshing was done with flails, most of it was done by slashing the wheat on a stone placed on a bench or on an upright barrel. The thresher would take a handful, or *dornán*, of wheat from a sheaf and, holding the bottom of it, bend it back and forth a number of times by drawing the top of it up and down against the edge of the bench. This made the wheat more flexible, but care had to be taken not to crack off the heads. The *dornán* of wheat was then threshed by slashing it repeatedly on the stone. Any short stalks of wheat in the *dornán* that did not reach the stone were pounded with a short stick known as a 'waspidel'.[34] The stone used had to have a 'square' edge as this provided more friction than a rounded one. The work was surprisingly hard. Initially, relatively small *dornán*s of wheat were taken to avoid swelling of the wrists, until they became conditioned to the work. Slashing provided better straw for thatching than flailing did, as it did not crack the straws. A sheaf contained about three *dornán*s, and each sheaf was rebound after it was threshed. In this way, the individual straws were kept running parallel with each other with the cut ends together, right through the entire process of reaping, binding into sheaves, threshing and re-binding until the straw was eventually placed in position on the roof. Up to the mid-1920s, practically all of the houses on Heir Island were thatched.

The unpredictable weather at this time of year ruled out the possibility of any lobster fishing along the shore, but many lobstermen braved the changeable sea conditions for a few more weeks to fish locally for mackerel or herring. From mid-October through

November and sometimes into December, many Heir Island lobsterboats fished herrings over a wide area of the bay, mooring their nets just after dark and hauling at dawn. The Sands and especially the Éanach, west of Sherkin Island, were particularly good herring grounds. In shallow harbour waters, the presence of herring was indicated by a type of seaweed, known as *corraí*, floating on the surface of the water, having been churned up by large shoals moving close to the bottom.

The autumn mackerel season extended from August to early December, with the shoals coming relatively close inshore which enabled small craft like lobsterboats to compete with the large drifters operating out of Baltimore. The Long Island boats, in particular, would always take to mackerel fishing at this time of year, fishing with drift-nets at night, north-west of Cape Clear.

Mackerel fishing at night in a small boat at that late stage of the year was particularly hazardous, especially as the boats were frequently laden down with large catches. Yet, one of the more unfortunate drownings associated with the fishery occurred in broad daylight in the relatively sheltered waters of Long Island Channel on a day that, for the most part, seems to have enjoyed relatively good weather.

On Saturday 5 December 1908, a violent *sí gaoithe* ripped its way through the townland of Colla on the north shore of Long Island Channel. The whirlwind, usually a feature of summertime weather and rarely occurring at that time of year, left a trail of havoc in its wake. As it raced haphazardly across the countryside, farmers paid homage to this 'fairy wind', urgently offering it handfuls of straw or hay or any other material that came to hand in the hope of appeasing its anger, lest it demand a greater sacrifice. As Denny 'Paddy' O'Sullivan unloaded a cart-load of furze, it roared across his haggart, upturned his cart, breaking one of the shafts, and headed towards the sea. No stranger to the perils of the sea himself, Den O'Sullivan turned to his wife Catherine, standing nearby, and lamented the fate of any crew whose boat might be unfortunate enough to cross the path of the *sí gaoithe*.

Little did Den realise how prophetic his words would prove to be, for sailing up Long Island Channel as he spoke was the lobsterboat *Water Lily*. On board was her hardy eighty-four-year-old skipper, Johnny 'Flor' O'Driscoll, and his eighteen-year-old son Jack. Having landed their catch of mackerel that morning at Baltimore, they were heading home to Long Island and were no doubt looking forward to what might well have been their first night's sleep in a week. Some distance ahead of them was John O'Donovan's boat of Gun Point, also heading westward.

As any sailor knows, two or more sailing boats heading in the same direction frequently results in a race. With their sails carefully trimmed in a steady, moderate breeze, and with no apparent danger of any sudden squalls, the O'Driscolls had the mainsheet of their gaff rig firmly 'cleated'. Possibly preoccupied with the boat ahead of them, they failed to notice the whirlwind rapidly approaching across the channel as it whipped up a quickly moving mist of foam and spray from the surface of the sea. Too late, the old man sensed the danger and shouted to his son to let go the mainsheet.

In seconds, it was all over. The *sí gaoithe* had come and gone on its merciless way and left Jack O'Driscoll lying on the floating pile of mackerel nets near the capsized boat. On Long Island, Florrie 'Batt' Whooley had seen the disaster take place and raced to the Céadán to launch his small punt. He rescued Jack O'Driscoll, but for the old man there was never any hope. As a young man, Johnny 'Florrie' O'Driscoll had survived the Great Famine, and for many years had outwitted the treacherous seas along the south coast. It was as if the elements had been patiently lying in wait for him and, finally, catching him off guard for an instant in the familiar waters of his own Long Island Channel, stealthily snatched his life away.

The spring mackerel season extended from about mid-March to June. In contrast with the autumn season, the mackerel shoals remained high out to sea, usually beyond the range of smaller boats. Some lobsterboats did, however, occasionally venture far out in search of spring mackerel. Thus, the night of Tuesday 11 May 1915 found the little twenty-five-and-a-half-foot lobsterboat *St Patrick* fishing among the large mackerel drifters off the Fastnet. On board were her skipper Mikey 'Mick Óg' O'Regan, his son Jackie and James O'Driscoll, all from Long Island. The previous Friday afternoon had witnessed one of the greatest maritime disasters ever to occur off the Irish coast when the *Lusitania*, pride of the Cunard Line, was torpedoed and sunk by a German U-Boat ten miles south of the Old Head of Kinsale. Over the following days and nights, many of the 1,200 bodies of the passengers and crew who were drowned drifted west along the coast together with masses of floating wreckage. The spring mackerel season was in full swing, and the mackerel fleet of Baltimore and Cape Clear played a large part in the grim task of recovering the bodies.

On that Tuesday night, just as they were about to shoot their nets, the crew of the *St Patrick* came upon a capsized lifeboat. Knowing that there was a possibility of bodies trapped inside, they sacrificed their night's fishing and settled down instead to a hard night's rowing with the lifeboat in tow. On reaching Schull Harbour the following morning, they could see that the lifeboat was marked '22A, *Lusitania*, Liverpool', and they reported their find to the coastguard. On righting the lifeboat, a most poignant scene met their eyes. The boat contained the bodies of a four-year-old boy and four women, one with her baby boy still clasped to her body by a lifebelt.

As well as herring fishing and mackerel fishing, some lobstermen, especially the Long Islanders, also found time for fishing with long-lines, known as spillers, and handlines which yielded a huge variety of fish including ling, cod, hake, conger eel, skate, halibut, turbot, plaice, sole, whiting, bream, wrasse, gurnet, pollock and coalfish. Coalfish are known locally as 'black pollock'. It is said that when an Eskimo looks at the world around him, he does not see simply snow, as would a person whose sense of perception has been conditioned by a more temperate environment, but several different types of snow, and that his language has a name for each variety he sees. Similarly, when a Roaringwater Bay fisherman looked at a catch of fish, he did not see simply a number of different species, as a more casual observer would, but distinguished within the more common species between fish of a slightly different size or of a slightly different colour, and again his language had a name for them all. A coalfish, for example, could be a *cuilig*, a *crothóg*, a *rabhlin* or a *gréasaí*,[35] depending on its size, a *cuilig* being the smallest and a *gréasaí* being the very largest. A *burd buí*[36] was a very small pollock of yellowish hue.

Around the end of November or sometimes as late as December, the lobsterboats were finally laid up for winter. On Heir Island, two days were set aside especially for the task, and the entire community co-operated in hauling up the boats, with messengers going from house to house to announce that the boats were about to be hauled. Organising themselves into groups, it would take the

Victory II, of the Aughadown fleet. Her crew, Jimmy Coughlan, Phil Dwyer and Eugene McCarthy, are thought to be putting their pots ashore at the end of the season. The mainsail is reefed and she is anchored by the stern.

islanders the full two days to remove the sails, rigging, masts, ballast and all equipment from the boats, and haul them up, with the aid of rollers, well above the high-water mark where they were propped in an upright position with their keels raised a few inches off the ground on flat stones and planks of wood. Not being under cover, the spiles were removed to allow rainwater to drain out. The boats were hauled up at several sheltered locations on the island including at the Guairneán, The Reen, near Burke's house and near the bridge. The *Flower of May* and the *Lily* from West Skeam were also hauled up on Heir Island, where both manpower and shelter were more readily available than on the angry shores of their own island.

Safe from winter gales and the strong tidal surges of Roaringwater Bay, the fleet remained until the following spring and the start of another season.

12 · Pot-making

The making of lobsterpots was a winter-time occupation. One season was the life expectancy of the fragile, wickerwork pots, and so they were replaced every year. Most households had an osier-bed producing osiers for basket-making, but these were too brittle for making lobsterpots. Instead, the stronger and tougher, wild willow withies[37] and twigs were used. The withies, or 'woodies' as they were termed locally, were cut and gathered on the mainland at several locations from the banks of the Ilen River to Gun Point, the lobstermen frequently walking miles inland in search of good-quality willow.

Skeaghanore and Greenmount, on the shores of Ballydehob Harbour, were convenient sources of withies for the Heir Island men. The hardy willow trees (the trees, themselves, were also referred to as woodies) that grew on the hills north of Ballydehob, however, produced withies of exceptional quality and were particularly sought after. In November and December, the months favoured for the task because of the low sap level in the willow, the crew of each lobsterboat would come to Ballydehob in their small lugsail yawls to cut the withies. They worked up to four miles inland, to the slopes of Bearna' Gaoithe, cutting the withies with reaping hooks and bringing them back to their boats in the evening. Occasionally, a pony and cart would be hired in the village, or an obliging farmer would volunteer the use of his horse and cart, but usually the men carried their bundles of withies on their backs. Each man could carry enough to produce about four pots. The crew of three, therefore, had to make at least two trips to Ballydehob to provide enough withies to make up a train of twenty-four to thirty pots.

The new moon in November was considered the optimum time to cut withies as it was held to mark the very end of nature's annual growth cycle. At that particular time, all growth is said to come to an absolute standstill before re-awakening with the waxing moon. Withies cut at this time were tough but pliable; and the sooner they were used after being cut, the better, as they quickly became brittle. Their pliability, however, could be preserved by immersing them in a stream or pond until such time as they were used.

The wickerwork lobsterpot was basically a cage-like contrivance. It had a circular base and narrowed slightly towards the top before curving in sharply to the pot entrance, or mouth. The size of the pots varied depending on the preferences of the individual pot-makers, and estimates as to the typical height of a pot vary from eighteen to twenty-four inches. The diameter of the base would have been several inches greater than the height. The mouth of the pot, which was eight or nine inches in diameter, extended down into the pot in the form of a wickerwork cylinder six to nine inches deep, and was referred to as the 'chimney'. No doubt it was the wickerwork chimneys which were a feature of many humble dwellings in the nineteenth century that prompted the use of this term.

Fig. 1

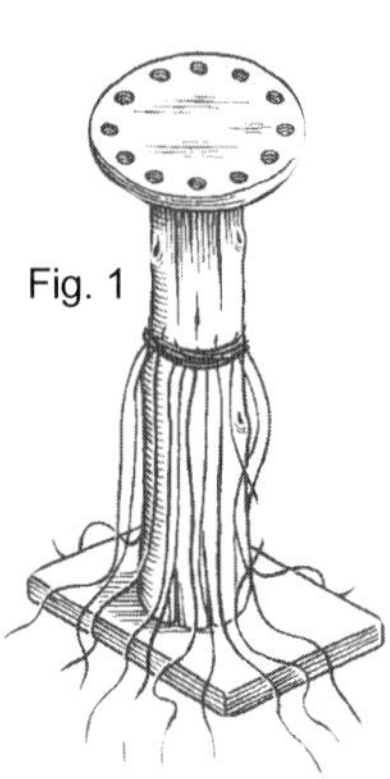

Illustration: Russell Barnett

The lobsterpots were made with the aid of a simple wooden mould (fig. 1).[38] This consisted of a wooden disc nailed in a horizontal position on to the top of a vertical length of wood, or leg. A series of holes was drilled through the disc in the form of a circle eight or nine inches in diameter. The leg, which was about three feet in height, was fashioned from any suitable piece of stout timber that came to hand, quite often from the branch of a tree. The mould was kept standing vertically by a short plank of wood nailed across the bottom of the leg on which stones were placed if necessary or, alternatively, by driving the leg into the ground. In later years, it was set in a block of concrete.

The number of holes in the wooden disc varied from eleven to fourteen, but most pot-makers seem to have used twelve. The first step in making a pot was to strip the withies of any small branches or twigs, and cut them to about four and a half feet in length. Twelve strong withies, referred to as 'rods', were then selected. The thick ends of these rods were inserted into the holes in the wooden disc, one in each of the twelve holes, so that they stood vertically on top of the mould. These vertical rods were referred to as 'standards'. Some of the light twigs which had been stripped from the withies were then woven horizontally in and out through these standards, all the way around, to form a solid wickerwork cylinder rising six to eight inches in height above the wooden disc. At this point, a further twenty-four, slightly lighter, standards were threaded vertically down through the wickerwork of the wall of the cylinder, one at either side of each of the original twelve. This left a total of thirty-six standards rising vertically from the wall of the cylinder (fig. 2). All thirty-six standards were then bent, at the top of the cylinder, outwards in a gradual curve and then steeply downwards to form the framework for the body of the pot, leaving the cylinder extending down inside the pot to form the chimney. The standards were held down in this position by pieces of string tied to the leg of the mould.

Some pot-makers then continued the wickerwork wall of the chimney out and around the top of the pot for some distance to avoid the risk of lobsters reaching the bait from outside the pot through the gaps between the standards (the bait would be tied to the inside of the chimney when fishing). A wale was then constructed which started at the mouth and spiralled down and around the body of the pot. This wale consisted of two strands of light twigs which were twisted around each other like the strands of a rope. Each strand was made up of bundles of two or three twigs treated as one. One strand passed at either side of each standard, and the strands were given 'half a twist' around each other between the standards, so that the strand that passed inside of one particular standard would pass outside of the next and vice versa. Basically, it was as if the standards were passing between the strands of a two-stranded rope. This rope-like feature, known by its Irish term *buinne*, held the standards in position. The pot-maker gauged the distance between the standards by the width of two of his fingers. The *buinne* made four or five revolutions around the pot as it spiralled downwards before levelling off to make a horizontal circular rim which was to be the bottom edge of the pot. Before the *buinne* was completed, a length of rope was tied in the form of a circle around the standards so that the strings which had held them in place could be released (fig. 3, *see over*). As the *buinne* was being completed, the diameter and position of the circle of rope was adjusted to regulate the shape of the pot and to allow the standards to spread out a little as they approached the bottom of the pot. When the *buinne* was completed, it held the pot to this shape and the rope was removed. At this point, a further *buinne* was constructed which made one revolution around the bottom edge of the pot. This *buinne* consisted of two strands, each of which was made up of two heavy withies. The strands of this *buinne* were not twisted between every standard. Taking three consecutive standards, the strands were given 'half a twist' before and after the first one but passed at the same side of the next two before being twisted again. As well as providing extra strength, this *buinne* was intended to protect the bottom edge of the pot from wear and tear.[39]

Illustration: Russell Barnett

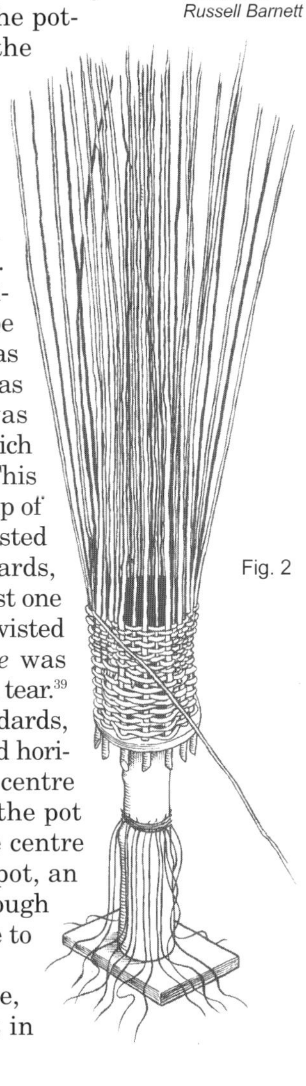

Fig. 2

The pot was then removed from the mould and the standards, which extended down beyond the *buinne*, were bent inward horizontally at the *buinne*, and cut to length to meet at the centre of the pot. This formed a framework for the bottom of the pot which consisted of thirty-six withies radiating from the centre like the spokes of a wheel. Starting at the edge of the pot, an ever-decreasing spiral of light twigs was interwoven through these radiating withies to give a solid, closely woven base to the finished pot (fig. 4 *see over*).

Two stones were tied to the floor of the pot on the inside, one at either side, to ballast it. Extra stones were put in

Fig. 3

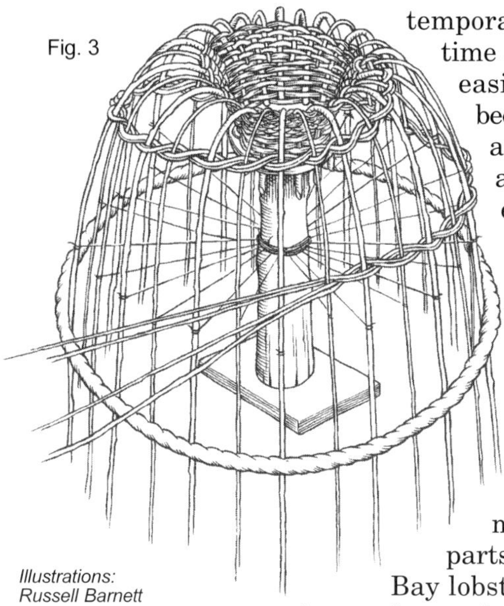

Illustrations: Russell Barnett

temporarily at the start of the season until such time as the pots became waterlogged and sank easily. It was believed that the pots did not become fully efficient at catching lobsters until a week's fishing had washed the residue of all natural juices out of the withies. At the end of the season, also, the pots tended to be less efficient. This was attributed to the fact that, by that time, the withies would have lost all of their bark and the pots would have become very bright and almost white in colour.

In England, this type of pot was baited by skewering the bait on pointed sticks which were positioned around the top of the pot on the inside.[40] It is said that this method of baiting was also used on some parts of the Kerry coast, but the Roaringwater Bay lobstermen preferred to use a piece of string, about a foot and a half in length, with a loop formed by a running knot at one end of it. The loop was positioned on the inside of the chimney with the other end of the string passing out through the chimney wall. The loop was closed on the bait by pulling the end of the string tightly and tying it to one of the standards on the outside, thus keeping the bait firmly in position against the inside of the chimney. Three pieces of bait were used, and the three bait strings were positioned at equal distances around the circumference of the chimney.

Generally, all three crew members of each individual boat co-operated in providing withies and in making pots. Some fishermen, like Florrie 'Batt' Whooley of Long Island, made pots for sale once their own requirements were fulfilled. A good pot-maker could make about three pots per day for which, around the year 1900, he would charge a total of half-a-crown.[41] It also seems that work was available for professional pot-makers who were not seemingly directly involved in fishing themselves. An interesting article, which appeared in the *Cork County Eagle* on 21 June 1912, details the results of a competition held for the best-made lobsterpots. The eventual winner of the competition was a Mr Jer. Collins from the inland district of Caheragh, Drimoleague:

Fig. 4

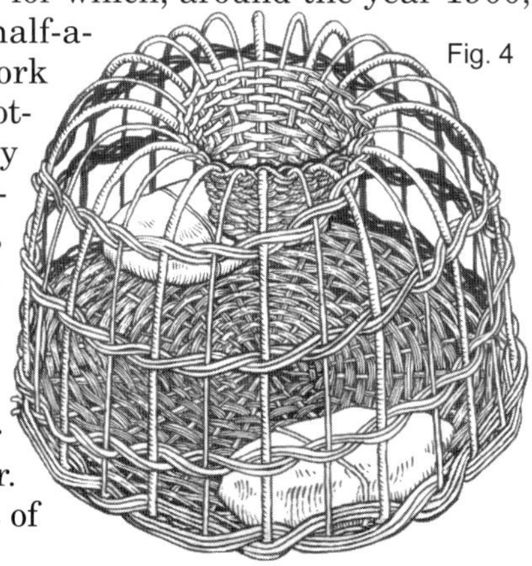

At a competition held recently for the best made lobster pots, 72 entered, the competitors being from all parts of Great Britain, and it is pleasing to know that Mr Jer. Collins, of Caheragh, beat all his competitors and secured the first prize of £7. The tests were very severe as to the capacity for holding the lobsters when once in, and Mr Collins' pots were the only ones which came through the tests. On the first night the 72 pots were put out, and next morning seven came out with 2 and 3 lobsters in them. The seven were immediately set, being again drawn at 8.30 at night, only three containing lobster; these three were again set and the following morning one was empty, thus leaving Collins of Caheragh, and Connor for the final. The two pots were then baited and placed where there were plenty of hungry lobsters, and it was not long before they were occupied. The pots were then disturbed and after waiting an hour were drawn, and Collins' pot was the only one which proved able to hold its catch. Mr Collins is to be congratulated on maintaining the honour of his native place in holding his own against such heavy odds.

13 · The Pre-Famine Lobster Fishery

Historical references to the early lobster fishery in Roaringwater Bay are very scant. Pococke, a tourist visiting the area in 1758, noted the large variety of fish landed at Crookhaven. Among them, he listed lobsters, crabs and crayfish which, he wrote, sold 'for pence a piece'.[42] It is not clear, however, how these crustaceans were caught. It is known that on many parts of the coast, lobsters and crabs were taken from clefts in the rocks at low water by means of a pole about six feet in length with an iron hook attached to the end of it.[43] When plentiful, they could also feature as a by-catch of shore sein nets, which were hauled onto the shore. This was a common fishing method at the time. Crayfish, on the other hand, usually occupy offshore reefs. Therefore, although they were a by-catch of trawling when that method became common early in the following century, they would be far less likely to be caught in a shore sein, and could not usually be caught by pole and hook. This would suggest the use of pots, or possibly even some kind of tangle nets[44] fished by boat. There is, however, no definite evidence to support this theory, and we cannot say for sure that boats set out to fish specifically for lobsters or crayfish in Roaringwater Bay at this time.

The earliest direct reference we have to Roaringwater Bay fishermen targeting lobsters specifically comes from Robert Wilson, who was appointed light-keeper in the lighthouse in Cape Clear in 1817. In his *Brief Description of Cape Clear*, published in 1832, he wrote that lobsters and crabs 'are taken by people from the mainland, in a kind of pot, made of three or four hoops covered with a net, and a hole in each end; about a dozen of these made fast together, with fish for bait, are sunk near the rocks and a floating mark to find them by.'[45]

One of the interesting things about Wilson's description of the lobsterpots used by Roaringwater Bay fishermen at that time is that it matches very closely a description by Wallop Brabazon in 1848 of the pots used in England. In his book, *The Deep Sea and Coast Fisheries of Ireland*, Brabazon described 'the English lobster trap' as being 'made of four or more hoops tied at equal distances, on spars about four feet long; there is a strong netting over this, with two entrances into it; one at either end.'[46] Brabazon claimed that this English pot was more efficient than the Irish 'basket' type pot. The 'basket' type pot, he informs us, was the one used 'all along the Irish coast'.[47] It would seem, therefore, that while the 'basket' type pot was the one generally used by Irish fishermen, the fishermen of Roaringwater Bay were using the English-type pot at this time. This English-type pot, as we shall see, provides a pre-Famine link between the lobsterboats of Roaringwater Bay and the waters to the east known to have been fished by the fleet in later times.

In the *First Report of the Commissioners of Inquiry into the State of the Irish Fisheries* of 1837, Lieutenant C.H. Kempthorne, CGO, Mill Cove, Glandore reported that 'Lobsters of most excellent

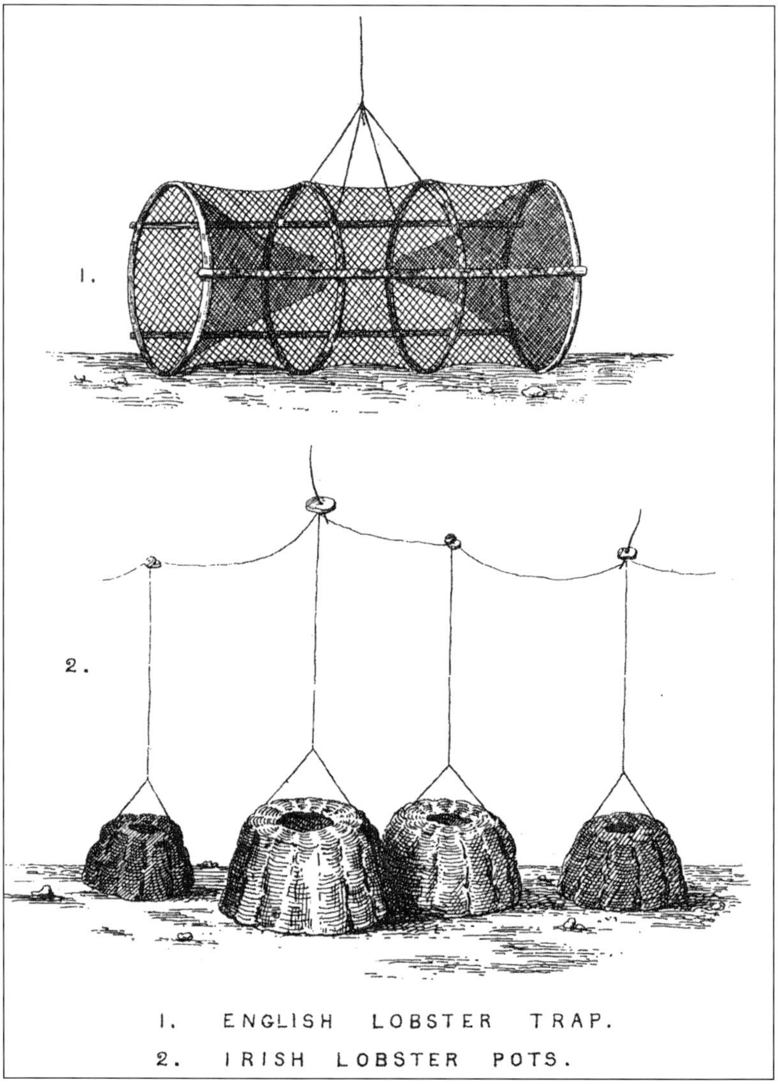

An illustration from Brabazon's *The Deep Sea and Coast Fisheries of Ireland* depicting 'the English lobster trap' as opposed to 'Irish lobsterpots'. The English-type pot was actually the type used by lobstermen in Roaringwater Bay before the Famine.

quality are found here, they are taken in pots by strangers, chiefly for the Cork market.'[48] A Mr Morris, also of Glandore, reported that 'From 15 to 20 boats visit the coast annually for the lobster fishery; only five or six belong to the district',[49] and, most interestingly, a Mr O'Donovan of Squince near the mouth of Glandore Harbour described the pots used as being 'formed of four wooden hoops, bound with laths, and covered with net.'[50] According to the folklore of Heir Island, lobsterboats from Roaringwater Bay had been making long fishing trips along the coast to the east for several generations. It is quite possible then, that the 'strangers' visiting the Glandore area in the 1830s, with their English-type pots, were the lobstermen of Roaringwater Bay.

The overall impression one gets from the 1837 *Inquiry* report is one of a drastically declining national fishery with lobster fishing, in particular, being one of its most neglected aspects. Although lobsters were a plentiful species on the Irish coast, they were scarce and expensive in Irish markets, and often were not available at all. Meanwhile, Norway was the main supplier of the English market. However, it does not seem to have been an inability or a reluctance on the part of Irish fishermen to fish lobsters that was the problem, but rather an inability to access these markets. Brabazon's book (published in 1848) claimed, for example, that there had been a successful lobster fishery in Clew Bay, Co. Mayo 'some years' earlier; yet while the 1837 *Inquiry* report noted the scarcity and high price of lobsters in Irish markets, Brabazon claimed that large female lobsters caught in Clew Bay were sold 'so cheap as 2 shillings, or 1 shilling and 6d per dozen', and fishermen could get no sale at all for male lobsters except in times of 'great scarcity'.[51] Poor communications and remoteness from markets were frequently mentioned with regard to Irish fisheries, and this would have particular significance for the transit of lobsters and crayfish which are highly perishable.

This difficulty in accessing markets would have had implications for the lobstermen of Roaringwater Bay also. The 'Cork market' was the main market, and apparently the only significant market for lobsters caught on the south coast at this time. A small amount was sold directly to the local gentry, but this would have been a very limited market. Lobsters caught in the Glandore area, for example, were 'sent to the Cork market'.[52] At Kinsale, however, which was within relatively easy striking distance of this market by horse and cart, there seems to have been a regular, organised trade in lobsters, with 'jobbers'[53] transporting them to Cork. No other port south of Cork City is mentioned in the 1837 *Inquiry* report in connection with the marketing of lobsters. Kinsale, then, would long have been an attraction for lobstermen from further west as a port at which they could easily market a large quantity of lobsters at a good price.

While there is no concrete evidence that the lobstermen of Roaringwater Bay did, in actual fact, avail of the marketing facilities at Kinsale or fish in that area as early as the 1830s, we do know from the 1837 *Inquiry* report that nearly twenty boats fished lobsters at Kinsale during the spring and summer; and while it is not specified whether or not any of these boats were visitors to the area, it was once again mentioned that the pots used were 'skeletons of wood hoops covered with netting'.[54] On the other hand, the pots used by fishermen at Youghal, further east along the coast, were 'of wickerwork (twigs)'.[55]

In the early 1840s, the Irish fishing industry, which had been in a state of decline, was showing signs of a revival only to be dashed by the devastating effects of the Famine of 1847. The industry went into free-fall when the mass emigration that followed in the Famine's wake was accompanied, for decades, by a steady decrease in the number of men and boats involved in fishing. The *Report on the*

Detail from a sketch by Robert O'Callaghan Newenham around 1820 gives us a very good picture of what pre-Famine lobsterboats may have looked like. The drawing shows Timoleague Friary at the head of Courtmacsherry Bay, a bay which was one of the favourite fishing grounds of the Roaringwater Bay fleet in later years. In the foreground is a moored boat with a towel erected. The mast is lowered and the towel seems to be draped over semi-circular hoops arched across the boat. In that period, a spritsail, or possibly a lugsail, would have been in use.

Coast and Deep Sea Fisheries of Ireland, by the Royal Commissioners on Irish Oyster Fisheries in 1870 gives the following account of the effects of the Famine on the industry:

> So broken down by want were thousands of the fishermen that they were physically incapable of proceeding to sea. Numerous instances occurred of the lives of those who made the attempt being endangered (some were actually lost), owing to their utter prostration from want of food having rendered them incapable of managing their boats.
>
> Multitudes had to part with their boats and gear for anything they could obtain to procure the means of appeasing the hunger of themselves and their families.
>
> At last nothing would be advanced on boats and materials, and a large amount of both rotted on the shores.
>
> A great number of those who survived the Famine were never afterwards in a position to purchase boats and gear, or repair what remained to them. Those whose friends sent for them emigrated. Several went to other pursuits and dispersed, year by year, the number following fishing becoming less.
>
> Thus it will be seen that the vessels and boats are now less than half, as compared with 1846, and the crews reduced to nearly a third.[56]

As with many aspects of the Famine, oral history sheds very little light on the fortunes of local West Cork fishing communities of the time. There is a folk-memory of the towelsail yawls fishing down east when the only food the lobstermen had to eat was a meagre supply of yellow meal (from imported maize) and whatever fish they could catch. This could also apply to several other periods of potato crop failure in the second half of the nineteenth century, however, and it is not possible to establish that local lobstermen succeeded in maintaining any significant degree of continuity in the fishery through the Famine years. On the contrary, the concentration of lobsterboats in Heir Island and its immediate vicinity forty years later, and the scarcity of such boats in other areas, would seem to indicate that Heir Island had been the focus of a new beginning in the fishery at some stage after the Famine. It is a strong possibility, however, that in fishing 'down east', these post-Famine lobstermen were knowingly continuing a pre-Famine tradition.

14 · Lobsters v. Mackerel

While the story of the pre-Famine lobsterboats and the extent of the fishery remains largely speculative, we do know at least (from Wilson's account) that lobsters were fished with boats and pots in Roaringwater Bay in that period.

By the 1870s, English fish merchants were sending boats over from Southampton to the Irish south coast to buy lobsters and crayfish from fishermen. Presumably, it was this new demand for lobsters that resuscitated the lobster fishery in the bay at a time when the fishing industry, in general, was still reeling from the after-effects of the Famine. Offering seasonal contracts to the fishermen, the captains of these English lobster cutters undertook to supply them with fishing gear and to buy their catch at a set price in return for which the fishermen guaranteed to supply them with lobsters and crayfish for the duration of the season. What was purported to be the contents of a letter from the captain of one of these English cutters to a local lobsterman is still preserved orally in the folklore of Roaringwater Bay: 'Charles Regan Sir, West Calf fisherman, Fish for me as you did before and I will supply you with pots, buoys and rope. Yours truly, Captain Hamilton.' The type of rope supplied was referred to locally as 'grass rope'. Presumably, this indicated that it was manufactured from sisal, which was frequently referred to as sisal 'grass'. As regards buoys, in the nineteenth century, they were generally made locally by fishermen from inflated pig skins and dog skins, or from inflated pig bladders covered with canvas and tarred. For the purposes of lobster fishing, however, they usually took the form of a large cork float, which was presumably the type supplied by the lobster buyers. (In later years, they were made up of a number of cork net-floats lashed together.) It seems to have been these English lobster buyers, also, who introduced the wickerwork lobster-pots to Roaringwater Bay. Although similar in shape to the pots described by Brabazon in 1848 as being the Irish 'basket' type lobster-pot, they were different in that the Irish pot had a solid wall like a basket, whereas these pots, as we have seen, were of a more open, cage-like construction. They were in widespread use on the south-east English coast, where they were known as the 'Cornish pot'.[57]

It is said that the Heir Islanders were among the first locals to learn how to make this type of pot. Later, when the Long Islanders took to lobstering, they would bring boatloads of withies to Heir Island, and employ the Heir Island men to make pots until such time as they became proficient in the art themselves. How it came about that the Heir Islanders became the pioneers of lobstering in Roaringwater Bay in this period is a question that remains unanswered. It is, however, possible to explain why most other fishing communities in the bay did not follow their lead. The fact is that shortly after the Heir Islanders took to lobstering, other fishermen became preoccupied with the development of another fishery, namely the mackerel fishery, which had a much higher profile economically, politically and socially, and was very actively promoted within the bay.

'One year . . . while fishing for herring, the Manx men at Kinsale reported that there was an abundance of mackerel, fine plump mackerel, on these south coasts.'[58] That was about 1860, and soon Kinsale became a base for a huge fleet of Manx, English, Scottish, French and Irish mackerel drifters. By 1880, a change in the migratory patterns of the shoals had brought the mackerel further south off the Irish coast, and with them to Baltimore came the fleet that hunted them. In the parish of Rath and the Islands, which included Baltimore, Sherkin Island and Cape Clear, the parish priest, Fr Davis, had already been campaigning to secure funding for local fishermen to buy the large, expensive boats, forty to sixty feet in length, which were capable of exploiting the highly lucrative mackerel fishery. In 1879, his efforts led to the English philanthropist, Baroness Burdett-Coutts, establishing an interest-free-loan fund to enable fishermen to import boats from the Isle of Man. By 1881, there were already forty-three of these large, sailing mackerel drifters registered locally,[59] with the fishermen of Cape Clear, in particular, eagerly taking to the fishery. In 1887, the Baltimore Fishery School, amid great public acclaim, was officially opened by the Baroness herself in the presence of many other dignitaries. The school was intended for the training of destitute boys in skills associated with the mackerel fishery. The opening of the school was reported in over seventy Irish and English newspapers, and Baltimore was depicted as a role model for the fishing communities of Ireland.[60] With vast amounts of mackerel being landed there by a huge international fleet, Baltimore had, in fact, become one of the most important mackerel fishing ports in Ireland. In 1892, for example, 270 mackerel drifters visited Baltimore.[61] With local boats included, a fleet of well over 300 boats anchored in the harbour was a regular occurrence. The boats had a crew of eight, and locals were employed not only on locally owned boats, but also on many of these visiting boats.

In the midst of this mackerel boom then, the captain of a lobster cutter visiting Baltimore would have found it very difficult to find

Two lobsterboats moored in the mackerel port of Baltimore in the 1890s. Heir Island and its environs were a stronghold of lobstering in a bay dominated by the mackerel fishery. In the background are barrels of cured mackerel destined for the American market, indicating that this photograph, from the Lawrence Collection, was taken during the autumn mackerel season; the spring mackerel were exported fresh on ice to the English market.

fishermen who were willing to take on a contract to fish lobsters were it not for that unique fleet of lobsterboats based in Heir Island and its environs. In the following decades, this corner of Roaringwater Bay remained a stronghold of the lobstering tradition while many other fishermen from all over the bay concentrated on the mackerel fishery and continued to invest in large mackerel drifters, importing them from the Isle of Man or, after 1900, having them built at the Fishery School boatyard in Baltimore.

This is not to say that all of the fishermen of Heir Island and its neighbouring communities were lobstermen. There were several imported Manx mackerel drifters in Aughadown, and Patrick O'Regan of West Skeam also had one. On Heir Island itself, the Burke family owned and fished a number of these boats over a long period of time; and Mike 'the Cloud' O'Neill was so nicknamed after a similar boat that he once owned, although he spent most of his years lobster fishing. Mike Cahalane, who also spent most of his life as a lobsterman, fished mackerel for some years when he had a ten-ton dandy,[62] the *St Nicholas*, built at Baltimore in 1918. Mike O'Driscoll, also of Heir Island, fished mackerel for many years with a similar boat known as the *Heber*. Lobster fishing, however, continued to be by far the dominant fishery of this area of the bay.

The situation on Long Island was somewhat different. Lobstering was, undoubtedly, a very important fishery for the Long Islanders but so also was mackerel fishing, especially in the autumn, when the shoals came relatively close inshore and within range of their small, open lobsterboats. There were no large drifters based on Long Island; but from about 1909 on, many of the Long Island lobstermen progressed to owning a mackerel yawl (many of which were imported second-hand from Cape Clear), and embraced mackerel fishing as their main source of income. These boats, ranging in size from twenty-eight feet to thirty-six feet in length, were much smaller and more affordable than the large, expensive drifters but larger and more suitable for mackerel fishing than the open lobsterboats.

A lobsterboat known as the *Oileán Garbh* represents an interesting, intermediate stage in the transition from lobsterboat to mackerel yawl by the Long Islanders. Owned by Connie 'Mick Óg' O'Regan, she had portable side decks which O'Regan fitted every year at the end of the lobster season when he set out to go 'fall fishing' for mackerel. Side decks, as well as foredecks, were a characteristic of mackerel yawls, and greatly improved the safety aspect

From 1909 on, many Long Island fishermen progressed from lobsterboats to mackerel yawls. This photograph, taken at Castletownbere in 1913, shows two Long Island lobsterboats, the S567 and the S948, tied up alongside the mackerel yawl, *Naomh Pádraig* (920S), owned by John Sheehan, Berehaven. The S567 was the *St Patrick*, owned by Denny 'Donnchadh' O'Donovan, and was built by the Pyburns of Toormore in 1909. The year after this photograph was taken, Denny Donnchadh's' son, 'Big Pat', had the *St Finbarr* built, a thirty-five-foot mackerel yawl. *Spillane Collection*

of a sailing boat laden down with mackerel. They also drained off much of the water dripping from the nets as they were being hauled. In a lobsterboat, however, where space was at a premium, side decks would have been considered awkward and in the way.

In 1909, Connie 'Mick Óg' went a step further when he had the mackerel yawl *Erin* built by Henry Skinner of Baltimore. Other Long Islanders were to follow his lead in subsequent years, with 'Big Pat' O'Donovan, Mikey 'Mick Óg' O'Regan, Johnny O'Sullivan, Con 'Hayes' O'Driscoll and eventually (in the 1940s) Flor 'Hayes' O'Driscoll all investing in mackerel yawls. On West Calf, Dan O'Regan's lobsterboat *Ellen* became popularly known as the *Small Dominic* after he had the big mackerel yawl *St Dominic* built in 1910.

15 · Illustrious Lobstermen

The busy mackerel port that was Baltimore in the 1890s provided the backdrop for one of the most dramatic sea rescues on the south coast. A central figure in the rescue was a fearless lobsterman by the name of Cahalane, from The Reen, Heir Island.

On the night of Monday 23 April 1894, 'the night of the big wind', a ferocious hurricane hit the Irish coast and lashed the sea into a fury, wreaking havoc on the large fleet of mackerel drifters fishing out of Baltimore and Crookhaven. A total of nine men lost their lives on the south coast that dreadful night. One of them, a man by the name of Quirke, suffered a particularly traumatic death when, having been washed overboard from the Manx boat, CT23, he became entangled in her nets. His fellow crewmen, unable to haul their gear in the face of the storm, were eventually forced to abandon their friend to his terrible fate. The other eight men lost comprised the entire crew of the *Florence* of Kilkeel, who had bravely fought through the night to reach the safety of Baltimore only to founder on Tuesday morning in the huge seas near the mouth of the harbour.

The barque *Christian Wilhelm* of Norway was also running for the safety of Baltimore that morning. With the storm still raging, she grounded on the rocks near Sandy Island off Sherkin Island's north shore. The chief officer and men of Baltimore coastguard station put out in the station's lifeboat to rescue the crew but could not get near enough in the big seas to take them off.[63] On Heir Island, Cahalane, who had seen the distress signals, gathered a crew of four good oarsmen and set out with them in his lobsterboat. Three times they tried to reach the barque's side but failed until, on their fourth attempt, they succeeded in taking off four crewmen and landing them on Sherkin Island. On returning for the remaining six crewmen, disaster struck the little lobsterboat. Just as all the crew of the *Christian Wilhelm*, including Captain Krendal, had scrambled on board Cahalane's boat, a huge sea threw the barque over on her side and down on top of the lobsterboat, sinking her. Miraculously, all those on board managed to climb back onto the *Christian Wilhelm*; but Cahalane's boat, estimated to have been worth fourteen pounds,[64] was gone for good.

The loss of his boat cannot have been Cahalane's main concern as he stood on the *Christian Wilhelm* with seas breaking over her. He knew, however, that his fellow islanders would not easily surrender him to the sea. The rescue had been closely watched from Heir Island to the west. Once it was realised what had happened, seven men set out, not for the *Christian Wilhelm*, but for Baltimore.

The several English fish merchants who bought mackerel at Baltimore all kept 'bumming boats' there. These boats were long, fast, highly manoeuvreable, six-oar rowing 'yawls'. The competition among the mackerel buyers was such that these boats were used to travel miles out to sea to buy the catch from mackerel boats before they ever reached port. The Heir Island men quickly boarded one of these boats, which belonged to Kelsall Brothers of Liverpool, and set out with her official cox'n, William Harton, in command, for the *Christian Wilhelm*. As they rowed out, Harton had a change of heart and refused to leave the harbour; but the Heir Island men took command of the vessel and put him ashore on Sherkin Island. Pat O'Driscoll, one of Cahalane's regular crewmen, then took the tiller. With his fellow Heir Islanders, Michael McCarthy, Michael O'Driscoll, Michael Cotter, Peter Murphy, Dan McCarthy and John Harte, he succeeded in bringing all those on the *Christian Wilhelm* to safety before she disintegrated in the still-raging storm.

The same crew rendered assistance to several Manx boats that same day, and were responsible for the rescue of the yacht *Eileen*, owned by Richard Sisk, which was in danger of being wrecked by a drifting hulk in Baltimore Harbour.[65] William Harton later claimed that it was not through lack of valour that he left the boat but feared that there may have been too many on board to take off all of those on the *Christian Wilhelm*.[66]

One of the brave young men who crewed Kelsall's yawl that day was, as we have seen, a man by the name of John Harte. Ten years later, the names Harte and Cahalane were to the fore once more in another dramatic incident. This time, unfortunately, the story was to have a more tragic ending.

It was Tuesday 13 December 1904, at about nine o'clock in the morning, when a small lugsail yawl set sail from Heir Island bound for Ballydehob in a good sailing breeze. On board were John Harte, his brother Dan, his father-in-law Denis O'Neill, Connie O'Driscoll and Nelius Cahalane. On reaching Ballydehob, they moored their boat near the Westland quay on the Skeaghanore side of the harbour. While two of the men walked into the village to do some shopping, the other three set to work cutting withies along the shore for the making of pots for the following lobster season.

That evening, the wind blew strong from the north-east, driving before it heavy showers of rain and sleet. The three men had spread out along the shore. At nightfall, the people of Skeaghanore could hear them call out as they tried to locate each other in the darkness. Eventually, all five men came together where their boat was moored, and set out on a dark night and in bad weather for their island home about three and a half miles away.

On that same evening, James Coughlan and his crew had been dredging *leoithín*.[67] Having landed their cargo at Kilkilleen, two miles across the bay from the mouth of Ballydehob Harbour, they sent for help to haul up their boat in the worsening weather conditions. Later, as they dried themselves around the fire at Coughlan's house in Kilkilleen, what seemed like distant cries for help were borne in from the bay on the wind. Coughlan and his crew ran to their boat a quarter of a mile away. Being short of help, it took them over an hour to get her afloat. Rowing out against the strong north-east gale, they searched in total darkness and heavy rain but found nobody. Hoping that they had mistaken the cries of seals on Illaunveigh for human cries, they returned to shore.

Sometime later, further west along the shore, an exhausted Cahalane dragged himself ashore and lay semi-conscious on the stony strand as the waves washed over him. Shortly after the Heir Island men left Ballydehob, a strong squall had hit their little boat and capsized her in the middle of the bay. Of the five men on board, only Cahalane could swim but they all managed to cling to their upturned craft. They called for help for some time in the desperate hope that someone, somewhere might hear them above the sound of the wind and the sea. As they drifted out the bay, Cahalane had to decide whether to stay with the boat in the hope that someone might eventually come to their rescue, or alternatively, make a daring attempt to swim ashore while he still had a chance of making it. Knowing that none of them could survive the bitter cold until morning, he cut off his heavy boots with his knife and struck out bravely for the south shore of the bay.

Gasping for breath between the steep, choppy waves, Cahalane swam blindly in the black night and driving sleet. It was later estimated that it took him three hours to swim the mile to the shore. Exhausted, he was barely able to crawl up the rocky strand. When he gained enough strength, he set off stumbling, barefooted along the rough strand for the nearest light, and eventually reached White Hall Lodge.

Cahalane, who was well known for his courage, ruggedness and unyielding determination, had once boasted that there was not enough water in the Fastnet Sound to drown him. The men he had left out in the bay knew that if anyone could make it to the shore, he would. The question on their minds was who would survive until he was able to get help to them?

Denis O'Neill, the oldest man among them, was the first to succumb to the cold and the battering of the waves. By the time a crew was organised and a boat had made its way around the headland from Whitehall Harbour and the upturned boat located, only

John Harte remained. He was unconscious, but he had held on so determinedly for so long that his fingers had become locked in a vice-like grip on the gunnel of the boat. It was only with great difficulty that his rescuers prized open his grip and got him on board their boat. There was no trace to be found of the other three men.

Harte and Cahalane were taken to Heir Island. Harte, still unconscious, was taken into Donal O'Neill's house, the nearest to the landing place. Straw was lit around him on the kitchen floor to warm him. It was obvious that he was close to death but it was the following evening before a doctor could be reached. The upturned yawl was recovered but there was still no sign of the bodies of the drowned men. A large fleet of boats dragged the bay for weeks, but to no avail. In the second week of January 1905, the bodies were washed ashore one by one.

As each body was recovered, it was taken back to Heir Island. On the day of the burial, the funeral procession made its way from the home of the dead man to Cuaisín na gCorp[68] where the coffin was placed for a moment, as was the tradition, on the coffin-shaped patch of rock near where the 'listrums'[69] grow in the form of a cross. Then, ignoring the steps further south, relatives took the coffin down the rocks to the waiting boat, as had been done with every coffin leaving the island long before and long after those same steps had been hewn out of the solid rock by a forgotten generation. As the boat was rowed away, with the corpse and priest on board, the other boats of the island were tied on to it in a long line one behind the other. As the procession of boats glided along the silent waters towards the mainland, the women gave expression to the sorrow of their people and their high-pitched keen floated out across the still waters of the bay.

Dan Harte, Denis O'Neill and Connie O'Driscoll, who had spent their lives fishing dangerous waters, often far from home, had drowned in their home waters of Roaringwater Bay. The undaunted Nelius Cahalane was to spend many more years battling the sea, and John Harte, after some weeks of serious illness, recovered and lived to save the lives of others.

Several years after this tragedy in another, now almost forgotten incident, the awful events of that night were brought forcefully back to John Harte. One night, while fishing for herring at the very same time of the year and in the very same waters in which he had so nearly lost his life, he thought he heard faint, distant cries for help. Terrible memories of that dreadful night came flooding back to him. Were these the cries of his dead brother, father-in-law and friend still echoing through the black night on Roaringwater Bay so many years after those same roaring waters had torn them, one by one, from his side?

Harte's crew tried to persuade him that what he had heard was the crying of seals. They were quickly silenced, however, when

Harte, abandoning his nets, headed his boat in the direction from which he had heard the cries coming. As they rowed through the darkness, they anxiously wondered what lay ahead.

What eventually appeared before them was an eerie image, as if projected through time from that night long ago, of an upturned yawl with two exhausted men clinging on desperately, too weak to call out any longer.

As in the earlier incident, the two men, Mike O'Neill and Charlie O'Neill, had been at Ballydehob cutting and gathering withies. While there, they had been informed that one of the village carpenters, Denis O'Driscoll, had completed a new donkey cart for one of them. At Ballydehob quay, they loaded the cart on to their yawl along with the withies. The cart was upside down across the gunnels of the boat. While sailing back to Heir Island, a squall of wind heeled them over and the shafts of the cart dragged in the water, overturning the boat. By the time John Harte found them, they were very near the limits of their strength and endurance.

This same John Harte was awarded a silver medal and a citation by the RNLI 'in recognition of his gallant conduct in pulling off in a boat on two occasions, and, at great personal risk, saving five persons from the Motor Fishing Boat "Thomas Joseph", of Dublin, which was wrecked in a strong north-west wind and a very heavy sea, on Sherkin Island, Co. Cork, on 10–11 November 1918.'[70] The *Thomas Joseph* had left Baltimore for Schull that evening on a trial trip to test out her new engine. When returning after dark, she ran aground on the Catalogue Rocks, quite near to where the *Christian Wilhelm* had grounded twenty-four years earlier. Of the eleven people on board, only the five rescued by John Harte and his companions survived. The other men involved in the rescue, Tim Murphy and Jerry McCarthy, were awarded bronze medals.

Among those lost in the *Thomas Joseph* tragedy was her skipper and owner, John Daly, who only the year previously had been one of four Cape Clear men to be presented with silver medals and testimonials of honour by the RNLI for their part in the rescue of forty-six of the forty-seven crew of the 6,000-ton *Nestorian* on Bullán Point, Cape Clear on 2 January 1917. Among the boats from all over the bay that went to the rescue of the *Nestorian*'s crew were several from Heir Island; and once again, Mike and Nelius Cahalane played a prominent role in the rescue. Remarkably, this was the third rescue over a period of five days in which the Cahalane brothers had been involved. On 29 December, they had assisted heroically in the rescue of sixteen crew members of the British steamer *Alondra*, which was wrecked on the Kedge Rocks off Baltimore. Two days later, with our other hero John Harte, and Dan Harte, they had boarded the Norwegian barque *Saurnaut*, which was in trouble off Toe Head, and brought her safely into Queenstown.

On Tuesday 21 January 1892, an incident occurred off West Calf Island that dismayed the entire seafaring community. On that day, the inhabitants of the island watched in disbelief as the captain of a schooner apparently abandoned a young boy, a member of his crew, to what seemed a certain death. Thanks to the courage and quick action of some of the islanders, the boy lived to tell the tale and made a sworn statement before a Mr Jagoe JP. The *Eagle and Cork County Advertiser* of 30 January 1892 carried his remarkable story as recounted by himself:

> My name is Isaac Baker; I am fifteen years of age, and a native of Windermere, in England. I served four years aboard a training ship in Wales, and was put on board the schooner *Petrel* at Liverpool, by the officer of the training ship. The *Petrel* brought a cargo of sleepers to Schull for the Skibbereen and Schull railway, and having discharged them, we were proceeding to sea. It was blowing a strong gale from the south-west. The *Petrel* missed 'staying' on the north side of the West Calf Island, which is situated four miles outside of Schull Harbour. Captain Griffith ordered me into the ship's boat, which was astern, presumably with the intention of preparing her for receiving the crew if the ship went down. The ship was driven against the rocks. After a while she turned off, and gathered headway. The small boat in which I was got under the starboard quarter, when a sea struck her, and flung me overboard. I went down under the ship and came to the surface after some time. I saw the captain walking over the quarter, and called to him to throw me a life-belt, I being unable to swim a stroke. The captain, however, took no heed of me, and I went down again, and on coming to the surface next, for the second time, I saw the ship had sailed away a good distance. I hailed again as loud as I could, and took off my cap and waved it, but there was no heed taken; so I gave myself up for lost. I then doubled myself up in the water and kept in my breath, as I was afraid of going down. I saw that I was drifting quickly eastward, and that I would be either dashed against the rocks or go down and be drowned. After a considerable time I heard shouting, and was shortly afterwards taken aboard a boat by the West Calf islanders, who carried me ashore, and, having attended to my wants, I am now completely recovered. I was over half-an-hour in the angry sea unable to swim a stroke, and only for the quick action of the islanders probably no more would ever be heard of me, as the captain did not even report the occurrence ashore.

One of the boy's rescuers, Dan O'Regan, corroborated his story:

> I in company with some other men on the West Calf Island saw the *Petrel* strike against the rocks on the northern side of our island about noon. After a while the vessel turned off and sailed away. The boy who was in the small boat astern was thrown overboard. We saw and heard him hailing the ship, but no notice was taken of him, as the ship sailed back for Schull Harbour. I, my brother Patrick, and Michael, and Hugh Driscoll, ran to the eastern end of the island, where our boats are kept, and put out as quickly as possible. There was a strong wind and a high sea. We left another boy ashore to direct us towards the drowning man. After a severe pull of about half an hour, we reached and picked him up in an exhausted condition. We took him home, gave him warm drinks and dry clothes, as we thought he was dying, but he recovered and was all right in the course of a few hours.

When the islanders brought Isaac Baker into Schull, he immediately reported the incident to the police and coastguard who visited the *Petrel* to investigate the matter. Captain Griffiths, however, persistently denied having seen the boy in the water, and the police did not arrest him. It was only when Griffiths arrived in Cardiff, Wales sometime later that he was placed under arrest. At Cardiff Quarter Sessions on 26 April 1892, he was sentenced to one month's imprisonment for neglect of duty by failing to send the ship's boat to the boy's rescue.[71] According to oral history, Griffiths was also deprived of his Master's Certificate, and the O'Regans travelled to Cardiff to give evidence at his trial.

On 21 May 1892, at Baltimore Fishery School, the O'Regan and O'Driscoll brothers were presented with medals which were awarded by the Board of Trade for their gallantry in saving the life of the young boy. The presentation was made by Captain Welch RM, and a large crowd of dignitaries, gentry and local residents attended the ceremony.[72] Dan and Patrick O'Regan were well-known lobstermen. Some years after this incident, Patrick, or 'Pad' as he was usually called, saved the life of a woman who fell into the sea while picking *duileasc*[73] near the Mizen Head. Both Dan and Pad were noted oarsmen and sailors, and Pad was in popular demand as a helmsman in yacht races at Schull and Baltimore regattas right into his eighties. He died in 1962 at the age of ninety-seven. Michael O'Driscoll lived in Heir Island in later years. He is remembered, far and near, as Mike 'the Heber'.

Patrick O'Regan, one of the recipients of a Board of Trade medal for gallantry in saving the life of Isaac Baker in 1892. He is pictured here in the late 1950s at the age of 94.

16 · The Merchants

As already mentioned, by the 1870s, English fish merchants were sending boats to the Irish south coast to buy lobsters and crayfish from fishermen. The boats of the English lobster buyers were generally referred to locally as 'lobster cutters'; but in actual fact, most of them were ketches of between forty and sixty tons gross, and from fifty-seven feet to eighty feet in length. They were specifically designed for transporting lobsters and crayfish, and had a large compartment in the hold sealed off from the rest of the boat but with several narrow slots in the planks of the hull, at both sides, which allowed the compartment to flood with seawater, thus keeping the lobsters and crayfish alive. This compartment was known as a 'well', and the lobsters and crayfish were dropped into it through a hatchway in the deck. The cutters made frequent trips across the channel to the Irish south coast, buying lobsters and crayfish at certain ports until they had a full cargo with which they returned to Southampton. The captains of these cutters dealt directly with the fishermen rather than with middlemen. On the Cork coast, they depended mainly on the Roaringwater Bay lobster fleet with which they rendezvoused at Kinsale and Baltimore. These were their two main ports of call, and seemingly, their only regular ports of call on the Cork coast. As we have already seen, pre-Famine lobstermen had depended mainly on the Cork market to sell their lobsters. The fact that large quantities of lobsters could now be sold at Baltimore meant that the lobstermen, as well as fishing to the east, could now also exploit the under-fished grounds to the north-west, returning to Baltimore to sell their lobsters.

The lobster cutters were such regular visitors to Baltimore that they could be depended upon to act as committee boats for the annual regatta held in August. In 1896, a lobster cutter owned by a Mr Scovell of Southampton was used for the purpose.[74] Later that year, on Wednesday 7 October, four lobster cutters were anchored amidst the autumn mackerel fleet in Baltimore Harbour. Bad weather had already claimed two mackerel boats that season, and with the onset of a fierce gale, the fleet ran up Church Strand creek for shelter and waited out the short-lived but severe storm, safely anchored. Two of the lobster cutters, the *Ceres* and the *Uzziel*, again owned by Mr Scovell, joined the mackerel fleet in the safety of Church Strand. The other two, the *Harriet* and the *Ada*, were driven onto the rocks at Ringarogy and wrecked.

The *Harriet*, belonging to a Mr Locke, of Warsash, Litchfield, Hampshire, was crewed by Captain William Distin, Joseph Panell, John Dickenson and Wyatt Bedbrook, all from Warsash. On board the *Ada*, owned by John Scovell, jnr. of Hamble, Hampshire, were Captain Pewter and three men, all of whom were rescued from the rigging of their sinking boat by a Mr C. Regan of Ringarogy, near whose house the wrecks occurred.[75]

Among the lobster cutters that called to Kinsale in 1904 was the *Bonnie Lass*, under the command of Captain Bowman. She arrived on 22 July, and on 23 July she sailed 'for the westward'. On 15 August, the *Victor* arrived from Southampton and 'sailed for Baltimore' the following day. At various dates from 3 June to 10 September in 1905, the *Ceres*, *Gesse*, *Lord Brougham*, *Victor*, *Bonnie Lass* and *Uzziel* were buying lobsters in Kinsale, and all left Kinsale sailing 'for the westward'.[76] The *Ceres* and *Uzziel*, as we have seen, had been survivors of the 1896 storm in Baltimore. The skipper of the *Lord Brougham*, Captain Coverly, lives on in the folk-memory of Roaringwater Bay because of the fact that he always paid for his lobsters in gold (sovereigns).

During World War I and up to 1920, Jim O'Connor of Sherkin Island skippered the fifty-seven-foot well-boat *Smelt*[77] for Frederick H. Keane, a fish merchant from Ardmore, Co. Waterford. In spite of the risks involved in wartime shipping, O'Connor collected lobsters and crayfish at Kinsale and Baltimore, and delivered them to Brest, France. After the war, he also collected lobsters at the Blasket Islands, where commercial lobster fishing had been introduced by an English boat around 1880.[78] Among his crew were Michael 'Mickey' McCarthy and Eugene O'Neill of Sherkin Island.

By 1921, O'Connor was skippering an English cutter, the *Mary Leak*, owned by Charles Saunders of Southampton. The registered manager of the *Mary Leak* was Bertie G.A. Scott, also of Southampton, who had been buying lobsters from the Roaringwater Bay fleet for some years. By this time, the *Mary Leak* had been fitted with two small auxiliary engines totalling twenty-four horsepower,[79] but still depended mainly on sail and carried a full ketch rig. In July 1921, O'Connor and his crew had a narrow escape when they were caught in a 90 m.p.h. hurricane *en route* from the Blaskets to Baltimore. Forced to run before the northerly wind eighty miles out to sea, they eventually fought their way north again to make landfall at the Seven Heads, over twenty-five sea miles east of Baltimore.

By this time, Irish entrepreneurs had been taking over the lobster trade from the English cutters. Around 1920, a Queenstown businessman, Clayton Love, set up a holding tank at the old lifeboat house in the harbour to buy lobsters and crayfish from the

Roaringwater Bay boats whenever they were in the area. He eventually became one of the main lobster buyers for the entire fleet. In the 1940s, he employed Danny Wilson to buy lobsters at Kinsale and transport them to Cobh (previously known as Queenstown) by truck. At Heir Island, Mike 'the Cloud' O'Neill kept holding tanks for him, and sent off the lobsters and crayfish by rail from Baltimore. The liners calling at Cobh provided Clayton Love with one of his main markets, but he also exported to Britain.

Con Burke of Heir Island was another major lobster buyer. He had two holding tanks moored off the east end of the island, north of the slip. These 'tanks' were basically semi-submerged, wooden cages in which the lobsters and crayfish were stored until they were shipped out. Burke regularly sent consignments of lobsters by rail from Baltimore, presumably to England which was still the main market for Irish lobsters. He also exported directly by sea to France and Belgium where crayfish, in particular, were in high demand. His Continental customers sent over 'French cutters' on a regular basis to collect the crayfish and lobsters at Heir Island. These French cutters were well-boats similar to the English cutters. Burke himself frequently travelled to France and Belgium to promote his business. Throughout the 1930s, he bought lobsters and crayfish every Monday morning at Kinsale and brought them back to his holding tanks at Heir Island with a motor boat called the *Surf*, originally a ship's lifeboat. *Horizon* is said to have been the name of another boat he once used in the business. Burke frequently stood a drink to the lobstermen once the business of buying and selling was completed. Bill Daly's pub in Kinsale was among the lobstermen's favourite hostelries, and it is said the largest order the proprietor ever took was from Con Burke for twenty-four pints of porter.

Another Heir Island man, Mikey 'Merchant' O'Driscoll, is also said to have been an agent for a French cutter at one stage, and his son Patrick also dealt in lobsters and crayfish for a short period. At Baltimore, Jer McSweeney and Denny 'Thade' O'Driscoll also exported to France. They employed Denis 'Timmish' O'Driscoll to collect lobsters every week at Long Island pier with his boat known as the *Buckie*.[80] In Schull, Dan O'Donovan bought lobsters and crayfish, and shipped them out by train from Schull railway station.

Some of the French 'cutters', like their English counterparts, were in actual fact ketches; but again, they were all generally referred to as 'lobster cutters'. A large fleet of these French boats had long fished lobsters and particularly crayfish off the south and west Irish coast. Up to sixty feet in length, they could carry three to five tons of fish in their wells which they landed back in France. Even in the 1930s and 1940s, many of these boats had not yet been fitted with auxiliary engines or wheel-houses. They were steered by tiller, aided by a tackle in strong wind.

Crookhaven was a favourite refuge for these French boats in bad weather, and down to the 1940s, a large fleet of them would occasionally be seen anchored in the harbour. This French presence led to one of the most important developments in the lobster market on the south coast, when Pierre Trehiou, from Paimpol in Brittany, opened the lobster pond at Crookhaven in 1927. Trehiou, a Breton nationalist and a respected sea-dog, down to his proverbial wooden leg, was known locally as Captain Perry Trehiou. As well as providing another market for the local fleet, especially for the Long Islanders who always fished to the west, Trehiou collected lobsters and crayfish regularly at Dursey Sound, at Portmagee and up along the west coast as far as Carrigaholt in Co. Clare with an old Brixham sailing trawler called the *Thrift*, in which he had fitted a well and an engine. Interestingly, it was from Trehiou that the native-Irish-speakers of the Blasket Islands learned how to count lobsters in French.

Trehiou stored the lobsters in his pond at Crookhaven until such time as he had a consignment large enough to ship out. The lobsters were packed in boxes with sawdust and ice, and transported by lorry to the railway station at Schull or Skibbereen to be exported to England. The crayfish were exported to customers in France, including L'Higrets, who sent over a cutter regularly to collect them. Some of these cutters had small auxiliary engines, but they depended mainly on sail. They were about sixty feet in length, and carried five tons of crayfish. When Trehiou's business was fully

Trehiou's lobster pond at Crookhaven, now in disuse. *Photograph: Cormac Levis*

established, he sent out one of these five-ton shipments about every three weeks. One of these cutters was the *Charles Madeliene* of Brest. She took a full cargo of crayfish from the pond in August 1928. She was described as the 'yacht cutter type', and, given favourable weather conditions, could do the voyage from Crookhaven to Brest in forty-eight hours.[81]

Trehiou also had three traditional Breton lobsterboats fishing out of Crookhaven. None of these boats had auxiliary power. The forty-foot *Eugenie* and the thirty-three-foot *Edmond* carried a full cutter rig. The forty-six-foot *Trimardeur* was a ketch.[82] The *Edmond* was re-named the *Star of the Sea* when all three boats were registered in the name of local man, Rickie Collins, who managed the pond, so that they could be fished legally within three miles of the Irish coast. Initially, each of these three boats was crewed by a French skipper and three locals, but later the skippers were also replaced by locals. They fished north along the coast as far as Kerry Head at the mouth of the Shannon, the Blasket Islands and Brandon Head being among their favourite fishing grounds. Each boat fished about sixty French barrel-type pots, rather than the withy pots. Initially, they practiced the French method of fishing the pots in pairs, the boat acting as a mother-ship to a punt which was used to shoot and haul two pots at a time and bring the catch back to the boat to be stored in the well. The Frenchmen always sculled rather than rowed these punts, even in the strongest wind. When the skippers of Trehiou's boats were replaced by locals, they reverted to the Irish method of fishing the pots in trains, usually with fifteen to twenty pots per train.

An old French punt at Crookhaven in 1998. The Frenchmen at Crookhaven used punts like this, worked from a mother-ship, to shoot and haul pots which they fished in pairs. Note the sculling notch in the transom. *Photograph: Cormac Levis*

While fishing in Bantry Bay in 1932, the *Star of the Sea* was wrecked on Roancarrig, fortunately without any loss of life. Unfortunately for Trehiou, this disaster was not to be the greatest loss his business was to suffer. Trehiou usually spent the entire lobster season in Ireland, returning to Brittany at the end of each season carrying a cargo of crayfish on the *Thrift*. At the end of the 1937 season, the *Thrift* was anchored in Crookhaven Harbour, ready to sail to France with a cargo of five tons of crayfish on board, when she dragged her anchor in a gale and was wrecked on Lamb's Island.

In spite of these setbacks, Trehiou continued in business in Crookhaven until the outbreak of World War II, when he closed down the pond and returned to France. In 1941, the pond was re-opened under the management of Hennessy's of Ballineen, who leased it from Trehiou and ran it throughout the 1940s.

17 · Decline and Boom

In the early decades of the twentieth century, emigration was taking its toll on the island population of Roaringwater Bay; but in an essay on their island home, children attending Heir Island school in the early 1930s were still able to report that there were thirty-two dwellings and twenty-five lobsterboats on the island. On some of the smaller islands, entire families had been migrating to the mainland. By the end of the 1930s, the Calf Islands had been deserted. However, many of the families continued fishing after moving to the mainland, and although the fleet was slowly decreasing in numbers and many of the surviving boats were in an aging and deteriorating condition, there was still a substantial fleet of lobsterboats fishing out of Roaringwater Bay in the 1930s. World War II, however, was to bring a slump in the market for delicacies like lobster and crayfish, and with it a severe reduction in the size of the fleet.

In the early years of the war, many lobsterboats were laid up for a prolonged period of time. But if the war brought a decrease in the demand for lobsters and crayfish, it also brought a huge increase in demand for many other species of fish which had a greater practical value when it came to feeding our hungry, war-stricken neighbour, Great Britain. Dogfish, for example, which were in plentiful supply and easily caught on long-lines in inshore waters and for which there had previously been no market, were suddenly making fourteen to fifteen shillings per stone. All around Roaringwater Bay, old boats that had not seen salt-water for many years were hauled down, patched up and pressed into action. The old lobsterboats now suddenly had a market value, and many of them were sold off to fishermen from other parts of the bay who repaired them, raised their gunnels to give them more freeboard, fitted them out with foredecks and engines, and used them for fishing dogfish. Some lobsterboats were also put into use at this time in supplying the increased demand for sea sand and seaweed resulting from the short supply of artificial fertilisers during the war. Before the end of the war, the Heir Island fleet had been reduced to less than ten boats. The owners of these boats, now receiving a much-reduced price for lobsters, supplemented their decreased income from lobstering with the good prices to be got for mackerel and herring.

If the 1940s was a period of decline in the lobster fishery, the following decade was to bring a dramatic and unprecedented boom in the entire industry. On Heir Island, Con Burke's business, which had been taken over by his son Pat, unfortunately failed to survive the war. The war also deprived Clayton Love of a valuable market when liners ceased calling at Cobh, but he continued to ship out lobsters to the UK; after the war, he also exported to Ostend, Belgium. In the early 1950s, Love extended his business, and his company, Clayton Love & Sons, bought lobsters and crayfish at ports from Kilmore Quay in Co. Waterford, right around the south coast and up the west to Clare, the Aran Islands, Galway and Donegal. Ballycotton, in particular, became another regular landing place for many Heir Island boats around this time. In 1956, the company opened a lobster pond on the shore at the Old Head of Kinsale under the management of Francie Healy. With holding tanks at Church Street in Dublin, Clayton Love & Sons exported to merchants in Germany, including Petersen and Esseman, and also to Spain and France. In the 1960s, they tapped the liner market once more, and set up tanks in Southampton specifically for this market.

At Schull, Danny O'Regan was involved in a more modest lobster-buying operation with two businessmen from Cork by the names of Walsh and Blower. In the early 1950s, O'Regan modified a natural rock pool at the east side of Schull Harbour to form a small lobster pond, and the lobsters were exported to France in a French well-boat. Walsh and Blower also had a French lobsterboat fishing along the south coast. She was skippered by Frenchman, Jean Marie Penzac, and was renamed the *Walblow*.[83]

At Crookhaven, Hennessy's, who had survived the difficult war years, gave up the business around 1950, and the pond was leased from Trehiou by two Frenchmen, Jean Oulhen and his nephew, Alain Oulhen. Shortly after this, Jean Oulhen retired, and his son, also Jean, took up the partnership with Alain. The Oulhens were both from Brittany; Jean, who was known locally as 'White John', was from Primel, and Alain, who had a company known as La Langouste, was from Roscoff.

In 1952, yet another Jean Oulhen arrived in Crookhaven. To distinguish him from his first cousin, he was known as 'Black John' Oulhen. 'Black John' went into business independently of his cousins, and employed a local man, Sonny O'Sullivan, to manage his three holding tanks which he moored off the pier in the harbour.

Like Pierre Trehiou in the 1930s, 'White John' and Alain Oulhen had three Breton lobsterboats fishing out of Crookhaven. They were all auxiliary-powered ketches with wells. The *Notre Dame d'Esperance* and the *Ariel* were both thirty-four feet and their wells could each carry half a ton of lobsters. The well of the forty-eight-foot *Fleur de France* could hold one ton.[84] As with Trehiou's boats, they were registered in the name of Rickie Collins who was still in

The wreck of the *St Ita*, formerly the *Fleur de France*, one of Oulhen's Breton lobster-boats, abandoned at Crookhaven in the mid-1960s. This is how she looked in 1989.
Photograph: Giles Norman

charge of the day-to-day running of the pond. The *Fleur de France* was renamed the *St Ita* when she was registered in Skibbereen. She was skippered by Dan O'Leary who initially had a French crew; but later, like the other two boats, her crew were all locals. The *Ariel* was skippered by Mike Daly of Cape Clear.

The Oulhens had agents buying lobsters along the south Cork coast. Jer McSweeney, John O'Driscoll, Donal Sheehy and Neily O'Regan of Baltimore, and Jimmy O'Reilly of Schull all acted as agents for one or other of the two companies. They collected lobsters by lorry at Crosshaven, Kinsale, Courtmacsherry, Castletownsend, Glandore and several other small harbours. An agent by the name of Sheehan was operating at Castletownbere, and Oulhens also bought lobsters at Dursey Sound and Garnish, right out at the end of the Beara Peninsula. From about 1955 on, they collected lobsters by lorry from as far away as Cleggan and Black Sod Bay, Co. Mayo. Jerry O'Mahony of Crookhaven collected lobsters for 'White John' and Alain at Grange, Co. Sligo, and lobsters were also sent down from Killybegs, Co. Donegal.

All of these lobsters and crayfish were exported by sea directly from Crookhaven. Among the French well-boats that took them to France for 'White John' and Alain Oulhen were the *Mayannic*, which was about eighty-five feet in length, the *Louton*, about sixty-five feet, and the *Rosco*, about seventy feet in length. The wells of these boats could each carry about five tons of lobsters and crayfish. Another boat, the *Plumarch*, about seventy-five feet in length, had a holding tank on board through which sea water was pumped, and she could carry ten tons. At the height of the lobster boom in the late 1950s, one of these boats would leave the pond every ten days from May to October or November with a full cargo. Added to this, 'Black John' Oulhen had his own well-boat, the *Fauvette*, carrying his lobsters and crayfish to France. She was about sixty-five feet in length and could also carry about five tons. All of these boats were powered by diesel engine but also carried steadying sail.

18 · The Winds of Change

The dramatic upsurge in the demand for lobsters on the south coast in the 1950s was to have a major impact on the character of the lobster fleet, on the lifestyle of the lobstermen themselves and on the general fishing routine of the boats.

As we have seen, there had been a sharp decrease in the size of the lobster fleet during the war years. In the late 1940s, the remaining boats were an old fleet with most of them dating from between 1892 and 1912. In 1932, Henry Skinner of Baltimore village had built the first auxiliary-powered lobsterboat, the *Atlanta*, for Philip O'Dwyer of Marsh, Aughadown. The following year, he built a similar boat, the *Con Florrie*, for Thomas Ganey of Crookhaven. These boats were built, more or less, along the lines of the old lobsterboats although they had greater rise of floor amidships. They carried the usual gaff mainsail and staysail along with their small auxiliary engines. Apart from these boats, very few concessions had been made to modernisation, and very few of the surviving lobsterboats had been fitted with engines when Jack Pyburn gave up his berth in the *Mary Kate* in 1947 to sign on board a merchant ship in Cardiff. Jack left behind a fleet which, though diminished in size, had changed little in appearance or in fishing technique since the previous century, and a community of fishermen whose unique, traditional lifestyle had been equally as enduring.

The winds of change, however, were on their way. In the late 1940s, the fishermen were getting sixteen shillings per dozen for lobsters and eighteen shillings to one pound per dozen for crayfish. In the early 1950s, with the increasing demand for lobsters and crayfish and the increasing competition between buyers, the price of lobsters rocketed to two pounds per dozen and crayfish to two pounds and ten shillings. Suddenly, lobstering had become a much more profitable pursuit, quickly leading to improved working and living conditions on the old lobsterboats, and quite often to an abandonment of them altogether by the younger generation of lobstermen in favour of larger, more comfortable boats. The personnel involved in the fishery also changed to a certain extent when fishermen from other islands and the mainland, many of whom had done little or no lobstering previously, were attracted by the high price of lobsters. These newcomers to the fishery also introduced boat types that were not traditionally associated with the fleet so that by the mid-1950s, the Roaringwater Bay lobsterboats were no longer the largely homogeneous fleet they had been up to the previous decade.

When Jack Pyburn returned from 'steamboating' in 1953, he rejoined a lobster fleet that had changed significantly during the six years he had been away. Any of the old lobsterboats that were still fishing had undergone a metamorphosis. Although they still carried at least a mainsail, for emergency use, they had all had engines fitted. The engines had dispensed with rowing which meant that a foredeck could be fitted to form a small fo'c's'le where originally the oarsmen were positioned, and the towel disappeared forever. The fire-pot was replaced by a Primus, and a coal-burning stove with a metal chimney coming up through the foredeck provided heat. The breaker of water was lashed to the foredeck and a rubber hose for siphoning out the water had replaced the leather spout. The gunnels were raised to give sitting headroom in the cuddy and a higher rail for carrying the French barrel-style pots that had replaced the old withy pots. The result was that the boats had lost much of their original gracefulness and beauty, and had become box-like and top-heavy in appearance.

Not only had the old boats changed in appearance, but their numbers were steadily declining as they grew older. Some of them were sold off for use as pleasure craft, many of them finding homes in the ports around Cork Harbour where they had become so well known over the years, and others eventually making their way onto the inland waterways and even to England. Others were sold to businessmen interested in cashing in on the lobster bonanza or to people taking to fishing on a part-time basis. Some rotted on the shore or came to a humiliating end as a roof on a hen-house.

As the old boats disappeared from the fleet, they were no longer replaced with traditional lobsterboats. Richard Pyburn of Heir Island, who died at a relatively young age, had probably built his last lobsterboat some time before 1910. One of his sons, also Richard, had inherited the trade but emigrated to the US. Flor O'Neill of Sherkin Island and Baltimore also ceased building lobsterboats shortly after this. Thomas Pyburn of Toormore had died in 1894. His sons, Dick and Tom, continued to build smaller boats down to the late 1930s but probably built their last lobsterboat around 1920. His other son, Jack, building at Crookhaven, would have built his last lobsterboat around the same time.[85] At Rathmore, Baltimore, 'Old Willie' Skinner died in 1921; and it was his son, also Willie,[86] who built what was almost definitely the last truly traditional sailing lobsterboat, the *St Margaret*. She was built in 1925 for John Harte of Heir Island. It was only in the 1950s, when the sudden boom in lobstering coincided with a rapidly declining fleet, that there was a great demand for boats in the lobster fleet once more. This time, however, the eager demand for boats resulted in the old lobsterboats being replaced with a random collection of second-hand, engine-driven boats of various types, shapes and sizes. In 1950, Neily Harte, whose father and grandfather had fished the *Hanorah*, bought a small motor-boat known as the *Barker*. The following year, he progressed to the thirty-one-foot *Sally Brown*, an English import. In 1953, Jack Pyburn, who had fished on the *Mary Kate* with Jim McCarthy, bought the *Mary Joseph*, a twenty-eight-foot sandboat. That same year, Timmy Cahalane, whose family had the *Fear Not* up to the early 1940s, bought the thirty-foot *Wild Wave*, another English import. In 1958, Neily Harte's brother Seán bought the *Ebinezer*, a thirty-six-foot mackerel drifter. The following year, John Harte, who had fished on the *Hanorah* with Dan Harte (no immediate relation), bought the old mackerel yawl *Colleen* and, in a reversal of the *Oileán Garbh* procedure sixty years earlier, cut away her side decks to make her more suitable for lobstering.

The further introduction of mackerel boats into the fleet by newcomers to the lobster fishery further heightened its haphazard appearance. Many of these boats were too big and deep to manoeuvre safely close inshore, and were not the most suitable for lobstering. Although some of them had occasionally engaged in lobster fishing in the French style (that is, with the aid of a punt to work the pots) after the American market for autumn mackerel collapsed around 1930, they had never been a very common sight among the lobsterboats. In actual fact, some of the mackerel men, considering themselves to be deep-water fishermen, had a certain disdain for the lobstermen, referring to them disparagingly as 'rock dodgers'.

The changing face of the lobster fleet in the 1950s is illustrated by this photograph taken at Kinsale in August 1954. Moored on the outside are two of the old lobsterboats. Although they still carry emergency sail and oars, they have had engines and fo'c's'les fitted and their gunnels raised. In the nearer of the two, the water breaker can be seen on the foredeck and the chimney of the stove. On the inside are two of the several mackerel yawls that joined the fleet at this time.

This all changed in the 1950s, when the mackerel men found themselves in trouble once more. Over the years, the spring mackerel shoals had been moving further and further offshore and out of the range of the mackerel yawls. By 1954, the shoals had moved out high to the west, six hours motoring from the Fastnet for a mackerel yawl. Most of the yawls had already given up mackerel fishing at this stage, and although larger boats continued fishing at these distances down to 1960, for the smaller mackerel boats the lobster boom had arrived just in time.

Not only were the boats of the fleet changing, and to a certain extent the personnel, but so, too, was the overall fishing pattern of the fleet. With the lobster buyers now collecting lobsters by lorry all along the coast, it was possible to land the catch at practically any small harbour or landing place from Ballycotton to Dursey Sound. Individual skippers now tended to make arrangements with certain buyers to have their lobsters and crayfish collected regularly at a certain port, and quickly fell into a routine of landing and fishing almost exclusively out of that port for the entire season, or even for several seasons on end. The result was that the fleet became dispersed along the coast with a few boats based at several harbours. This had the knock-on effect of encouraging some fishermen into the fishery at some of these harbours, who had never previously engaged in lobstering. There were now more boats than ever before fishing lobsters on the south coast; but although Kinsale and Cobh remained favourite haunts for many of the Roaringwater

This photograph was also taken at Kinsale and probably on the same day as the previous one. Moored on the outside is the *Sally Brown*, an English import, which joined the fleet when she was bought by Neily Harte in 1951. Inside of her is one of the old lobsterboats with raised gunnels and fo'c's'le.

Bay lobstermen, long gone was the time when Sunday mornings could see up to seventeen of their boats tied up in each of these harbours.

The lifestyle of the lobstermen also changed significantly at this time. With a much-improved return for their labour, the working week eventually became slightly shorter with the boats landing and selling their catches on Saturday afternoon rather than taking advantage of Saturday night's fishing, as they used to, and coming into port on Sunday morning. As regards diet, each boat stocked up with supplies on Monday morning before putting to sea, and boiled bacon and fried rashers and sausages replaced fish on the menu, with the lobstermen seldom bothering to do any line fishing. 'Shop bread' also came into vogue, and baking on board was a thing of the past. The sharing out of the week's income was now also calculated differently. Previously, the week's spoils were divided into four shares with two shares going to the skipper, one for his own labour and one for the boat, and one share each to the two crewmen. Now a five-share system was in operation with the skipper getting three shares, one share for his labour, one for the boat and one for the engine. That was after the expenses incurred in fuel and supplies were deducted.

Danny 'Pad' O'Regan on board the *Mary* at Schull in 1952. In the 'holt' are some of the French barrel pots that had replaced the old withy pots.

In the whole area of seafaring expertise, the traditional skills of sailing and rowing were being replaced by a working knowledge of the internal combustion engine and by the modern, though not inconsiderable, skills involved in fishing a boat under engine power. For the first time since the dawn of history, an entire generation of fishermen was growing up on the south coast with very little practical experience of the finer points of sailing. The art of traditional pot-making was also being lost. The new French barrel pots were much more efficient than the withy pots at catching crayfish, which were more valuable than lobsters, and they quickly replaced the old pots completely. The raw materials for the French pots, narrow branches split into laths, were imported from France by the lobster buyers and sold to the fishermen, doing away with the annual quest for withies. The actual construction of the pots was simpler and far less time-consuming than that of the withy pots.

It involved constructing a cylinder of laths by nailing them onto hoops and closing the ends with netting. In actual fact, these French pots had a certain similarity to the 'English type' pots once fished in the bay. The method of tying the pots into a train also changed. Rather than connecting the pots by means of an individual length of rope between every two pots, each pot was now tied to a single long rope that stretched the full length of the train. The pots were tied to this long rope with short pieces of rope, about four or five feet in length, known as 'stoppers'. This method of mounting the pots saved them from being strained when hauling, the long rope taking all the pressure. This was particularly important when mechanical pot-haulers, powered by the engine, were gradually introduced in the late 1950s.

The 1960s brought their own changes. An Bórd Iascaigh Mhara, which had taken over the old boatyard of the Fishery School at Baltimore, started producing its series of thirty-two-foot *béal* boats, so called because the name of each boat built was to start with the word *Béal*. The yard also started producing a series of twenty-six-foot, clinker-built, double-ended boats. The clinker construction had always been shunned by fishermen on the south coast but was now gradually accepted. With Skinner's yard at Baltimore and Paddy Hegarty of Oldcourt also building new boats, yet another wave of change came over the lobster fleet with newly built, modern boats replacing the old mackerel yawls and other craft that had originally replaced many of the old lobsterboats in the previous decade.

The half-decked *béal* boats had comfortable accommodation and wheel-houses; and having a high rail, they could carry sixty-six French barrel pots which were fished in three trains of twenty-two pots each. Fitted with pot-haulers to ease the work-load of the fishermen and with echo sounders to search out new grounds in deep water, they were the state-of-the-art lobsterboat of the 1960s. In 1962, Jack Pyburn replaced the *Mary Joseph* with the *Béal Bán*, and the following year, Seán Harte replaced the *Ebinezer* with the *Béal Boirne*. In 1966, John Harte sold the *Colleen* and had the *Béal Ceare* built. Alongside these modern boats, a few of the old lobsterboats still survived, including Neily O'Donovan's *Pride of Toe Head*; but with new, reliable, diesel engines replacing the old petrol-paraffin engines, even the emergency sails disappeared. The last days of the sailing fishing boats of the south coast were well and truly over.

The boats continued to fish as far east as Ardmore and even further, with some of the new boats, like the *Béal Bán*, fishing as far as Helvic Head. As a matter of fact, with some of the 'French' boats at Crookhaven still fishing up the west as far as Kerry Head, the lobstermen of Roaringwater Bay, for a few years in the early 1960s, were covering a longer stretch of coastline than ever before: some 180 nautical miles.

By 1957, lobsters had reached the staggering price of three pounds per dozen. From around 1960 on, they were bought by weight, an indication that they were becoming an even more precious commodity as heavy fishing was starting to take its toll on stocks. At this time also, news was reaching the south coast that fishermen on other parts of the Irish coast were earning a lot of money drift netting at sea for salmon. Around 1964, some of the lobstermen tried out this new fishery, and it proved so successful that it very quickly surpassed lobstering as the main fishery of the small boats of the south coast. For most fishermen, lobstering was now only resorted to after the salmon season finished at the end of August. The late 1960s brought yet another lucrative new fishery to small boats when shrimp pots were introduced from France. Roaringwater Bay itself, where shrimps had been trawled under sail in the old days, proved to be the richest shrimp-fishing grounds on the south coast. With the shrimp season starting immediately after the end of the salmon season, shrimping was an ideal combination with salmon fishing, and by the early 1970s, lobstering was further relegated.

The *Béal Boirne*, a third generation lobsterboat. The *Béal Boirne* was built in 1963 for Seán Harte who had previously fished the *Ebinezer* (originally a mackerel drifter), and whose father and grandfather had fished the *Hanorah*. The *béal* boats were the state-of-the-art lobsterboats of the 1960s.
Photograph: Kevin O'Farrell

19 · MEMORIES

Many Roaringwater Bay boats, now fishing salmon, still concentrated their efforts to the east where large numbers of salmon were homing on the mouths of the large rivers like the Bandon, the Lee and the Blackwater. Down to the late 1970s and early 1980s, one would still occasionally come across a boat from the bay lobster fishing out of harbours like Courtmacsherry and Castletownsend. But the days of the Roaringwater Bay lobster fleet were long gone. For a young generation of fishermen, life on the old sailing lobster-boats was no more than a folk-memory passed down in the haunting tales of the older fishermen, and it was while lobster fishing on one of those boats in the late 1970s that I first listened to the enthralling story of the towelsail yawls.

For more and more people, island life was also becoming a memory. As previously mentioned, the Calf Islands had already been deserted by the end of the 1930s. By the mid-1940s, West Skeam had been vacated, and by the end of the 1950s, so also had East Skeam. Around 1950, three Heir Island families accepted farms from the Land Commission in other parts of Co. Cork. These three families were the first of many to leave the island in this period, for in spite of the lucrative fishing of the '50s, '60s and '70s, or, perhaps, partially as a result of it, the population of Heir Island and Long Island declined dramatically. Many families now had the wherewithal to move to the comforts and conveniences of the mainland while retaining their property on the islands. Others sold their houses to people from other walks of life and from other places and for whom the sheer beauty and solitude of the islands were so attractive. By 1981, the population of Heir Island was thirty-five and that of Long Island thirty-eight.

At Crookhaven, the *Notre Dame d'Esperence* had been blown ashore and wrecked in the harbour during a storm in 1962. By 1967, the *St Ita* had ceased fishing and was abandoned in a creek nearby, where her sad remains can still be seen. Around the same time, the *Ariel* was taken to Baltimore to be renovated but was found to be beyond repair; she remained unused for some years at Church Strand before being bought by an optimist with the good intentions of making her seaworthy again.

In 1963, 'Black John' Oulhen wound down his own business and joined forces with 'White John' and Alain at the pond. In 1976, the Oulhens sold the business to a partnership which included Pat Doherty of Donegal and a Mr Samuelson. The business survived for only two years under this management, and the pond finally closed down in 1978. At the Old Head of Kinsale, Francie Healy, who had managed Clayton Love's pond, went into business on his own in 1966. In the early 1970s, he bought the pond from Love and continued to export to France, Holland, Belgium and Germany but eventually went out of business in the early 1980s.

Around 1984, after a long and eventful life, the *Hanorah* was eventually laid to rest at the Mill Cove in Schull Harbour, where for many years I watched her gradual decline. One day, in the summer of 1999, however, I visited the Mill Cove not to lament the slow death of a once proud and beautiful boat, but to witness her resurrection. On that fateful day, 106 years after she was built, the *Hanorah* was raised from the slime by Nigel Towse of Sherkin Island and Liam Hegarty of Oldcourt Boatyard. Relieved of her high, cumbersome foredeck and raised gunnels, ugly mementos of her desperate struggle to cling to life as a fishing boat in her later years, her inherent grace and beauty were released once more and emerged through the grime in defiance of her dilapidated condition.

On a fine summer's evening in 1999, 106 years after she was built, the *Hanorah* is floated from her mud berth at the Mill in Schull Harbour.
Photograph: Kevin O'Farrell

Nigel Towse, her new owner, is currently completing the painstaking and faithful restoration of *Hanorah* to her former glory. With much of her original planking and frames intact, thanks to Dan Harte's legendary diligence in maintaining his boat, methodical removal of layers of mud, peeling paint and tar revealed many interesting constructional details, down to the oak treenails in her frames and patches of what seems to be her original grey and green paint. Meanwhile, Gubby Williams of Heir Island Boat Works is undertaking the restoration of the *Aileen*, a boat built by the Pyburns of Toormore in 1910, while Mary Jordan and Colm Ó Cuileannáin of Baltimore have taken delivery of the *Fionn* from Liam Hegarty's yard. Built to the graceful lines of the *Hanorah*, the *Fionn* has been the object of widespread admiration among a community whose growing awareness and appreciation of its maritime heritage may yet lead to the day when a fleet of these boats will once again ply the bays and inlets of the south coast.

The way of life associated with the towelsail yawls, however, is past and gone forever. The sheer hardship of life on an open boat at the mercy of the elements, the hard work, the loneliness of long weeks away from home and family, the terrifying storms, the tragic drownings, the daring rescues, the joy of the safe return and glad homecomings, the excited anticipation of the 'first tripper', the relief brought by a rising wind after hours of rowing, the camaraderie and the happy gatherings at Kinsale and Cobh on Sunday mornings, the lifestyle of entire island communities whose yearly pattern of life and work revolved around the coming and going of the towelsail yawls, all of these things belong now to another world, a world that a few of us wistfully remember and most of us can only try to imagine.

Nigel and *Hanorah*. The painstaking reconstruction of the *Hanorah*, a labour of love for her new owner, Nigel Towse, begins with the fitting of a new transom stern and stem-post moulded from the originals. *Photograph: Kevin O'Farrell*

(*above*) A detail from the *Hanorah* showing one of the original treenails in her ground timbers.
Photograph: Kevin O'Farrell

(*left*) The beautiful 'wine glass' transom stern of the *Hanorah*.
Photograph: Kevin O'Farrell

(*below*) Moulds being taken from the *Hanorah* for the building of the *Fionn*.
Photograph: Kevin O'Farrell

Shipwright Fachtna O'Sullivan fits the levelling strake of the *Fionn* while Colm Ó Cuileannáin, joint owner-to-be, wants no interruptions. *Photograph: Mary Jordan*

Time Warp. The *Fionn*, built to the lines of the *Hanorah* at Hegarty's Boatyard, Oldcourt in 2000 for Mary Jordan and Colm Ó Cuileannáin of Baltimore. Although built for pleasure purposes with substantial decking, the similarity between this photograph and the one on p. 14, taken around 1925, is remarkable.
Photograph: Nigel Towse

In January 2001, the *Fionn* became the first lobsterboat to 'cross the Lough' (Roaringwater Bay) under sail in fifty years. *Photograph: Kevin O'Farrell*

Two traditional West Cork boat types side by side once more: the *Fionn* (left) in the company of two mackerel yawls, Nigel Towse' *An Rún* (right) and Brian Ormond's *Macalla* (centre). All three boats were built at Hegarty's in recent years. *Photograph: Nigel Towse*

20 · Lobsterboats and Their Owners

The following histories of individual boats have been compiled by piecing together information collected from several different informants and combining it with whatever information, if any, there is to be found on each vessel in the *Register of Sea Fishing Boats* for the Port of Skibbereen.

Although registration numbers were usually inscribed on the boats themselves, the names were not. This sometimes led, over a period of time, to boats becoming known by names other than those under which they were originally registered. In a few cases, the same boat was known by different names at different periods or even on different parts of the coast. In the following inventory, all boat names, registered and non-registered, are listed in alphabetical order. The history of each boat is given under its registered name, and entries under alternative names simply refer the reader to the registered name.

The amount of information supplied in the *Register* varies from boat to boat. The registered name of the boat, its owner, place and date of building, dimensions, tonnage and rig are often given. In some cases, however, not all of this information is recorded. In such instances, one has to rely on the oral tradition to fill in the blanks and also to provide the names of the builders, which are not given in the *Register*. More often than not, under the heading 'Mode of Fishing', one finds the entry 'nets and lines'. This seems to have been a standard and consistently inaccurate entry made, as a matter of routine, by some officials recording the registrations. There are also some obvious inaccuracies in the registered dimensions of some of the boats, and some blatant errors in the calculation of tonnage. It should be noted that registered dimensions are given in feet and tenths of a foot.

It is usually possible to trace some of the changes in ownership of each boat from the *Register*; this record, however, is also incomplete in many cases as some owners neglected to re-register their boat in their own name when purchasing it. In other instances, the boats continued to change hands long after their registrations were closed. In most of the following histories, the names of the non-registered owners are also given. The entire matter is further complicated by the fact that some people, who were otherwise acknowledged as the owners of a particular boat, actually registered the boat in their father's name or, as in one instance, in their mother's name. In any such cases, it is the names of those popularly acknowledged as owners, rather than the registered owners, that are given here, unless otherwise stated.

As regards the boats built by Flor O'Neill, many informants were of the opinion that they were built on O'Neill's native Sherkin Island. A few were adamant, however, that all the known lobsterboats that he built were built when he lived at Baltimore, and this is borne out by the *Register* which gives Baltimore as the place of building for all of the boats traditionally attributed to him.

It should be noted that there were numerous boats of the lobsterboat design in Roaringwater Bay that were not generally used for lobstering. Many of the boats that qualified for inclusion in the lobsterboat class at regattas, for example, never actually fished lobsters. The following inventory is, of course, restricted to those boats that were actually part of the lobster fleet. References to lobsterboat races at Baltimore, Schull and Crookhaven regattas are taken from reports in the *Eagle and Cork County Advertiser* and the *Southern Star*.

Aileen S726

Builder:	The Pyburns, Toormore	
Date built:	1910	
Date Registered:	15/8/1928	
Registered dimensions:	Length of keel:	24.5 ft.
	Length overall:	28.0 ft.
	Breadth:	7.5 ft.
	Depth:	3.5 ft.
	Tonnage:	3.31
Rig:	Gaff mainsail and foresail	

At some unknown date, obviously after a major overhaul was carried out on the boat, her registered dimensions were changed to the following:

Length of keel:	25.0 ft.
Length overall:	28.0 ft.
Breadth:	7.7 ft.
Depth:	4.75 ft.
Tonnage:	4.59

Although built in 1910, this boat did not join the lobster fleet until the 1940s. She was originally built for Johnny O'Neill of Sherkin Island, who mainly used her to transport supplies, twice per week, from Skibbereen for his busy shop on the island. It has been suggested that the O'Connor family of Sherkin Island also had a share in her at some stage. As early as 1916, an engine had been installed in her. At that time, she was one of the very few small work boats in the area with an engine, enabling O'Neill to outdo all competition in salvaging shell casings from the *Nestorian*, which was wrecked on Cape Clear on the night of 2 January 1917. O'Neill, a coal importer, also used the *Aileen* to tow coal lighters when discharging coal from colliers delivering to him at Cuinne in Sherkin Island. His son, Michael O'Neill of Church Cross, later put her to similar use on the Ilen River.

Around 1940, Michael O'Neill had the *Aileen* rebuilt by Denis Desmond of Conomar. Although maintaining her overall length at twenty-eight feet, Desmond is said to have rebuilt her as a much more substantial craft than her original lobsterboat design. Presumably, it was Desmond's alterations that accounted for the changes in her registered dimensions and tonnage noted above. At this time also, her 12/14 h.p. Kelvin petrol-paraffin engine was replaced with a 13/15 h.p. engine of the same make. A bridge deck was built over the engine and also a foredeck.

Sometime after this, the *Aileen* joined the lobster fleet, and she was fished for O'Neill by various locals over a number of years, including some Heir Island men. In 1956, O'Neill sold her to Clayton

Love & Sons who employed Con Minihane of Garrettstown, Kinsale and two others from the locality to fish her. Minihane was originally from Heir Island where he had owned and fished the *Irish Leader* and, later, the *Safe Return*. At the end of the 1956 season, the *Aileen* was badly damaged during a storm while moored at Love's lobster pond at the Old Head of Kinsale. In November of that year, John Fitzgerald of Heir Island took over as skipper and had her towed back to Baltimore where she was repaired at Skinner's. John fished her for Clayton Love & Sons from 1957 to 1962, and always carried a lugsail as a standby. Another Heir Islander, Patsy Murphy, fished her for one more season before Love's sold her to Michael Daly of Cape Clear in 1965.

At this stage, Skinner's carried out further repairs to the *Aileen* including the fitting of a new keel. A dodger was also fitted and a 22 h.p. Saab diesel engine, fitted up forward, replaced the Kelvin. As with many of the old boats, it was found that she had been planked with pitch pine. Michael Daly fished her for lobsters up to 1981.

In 1984, the *Aileen* again changed hands when she was bought by John O'Donoghue of Bantry, and underwent even further repairs and alterations, this time at Paddy Hegarty's Boatyard, Oldcourt. O'Donoghue fished her for scallops, lobsters and salmon until she was sold to Brian Wall, Gerahies in 1994. In 1999, she was bought by Gubby Williams of Heir Island Boat Works, who is currently in the process of restoring her to her original sailing condition. She is said to have been a good sailer.

The *Aileen* at Bantry in 1998.
Photograph: Cormac Levis

Annie

The *Annie* was owned by Nelius Cahalane, The Reen, Heir Island in the 1890s. She won the lobsterboat race at Baltimore Regatta in 1896. She would have had the usual gaff rig of the Heir Island boats.

Atlanta **S193** also **D195**

Builder:	Henry Skinner, Baltimore	
Date built:	1932	
Date registered:	16/7/1932 (S193)	
Registered dimensions:	Length of keel	26.0 ft.
	Length overall	27.0 ft.
	Breadth	7.3 ft.
	Depth	2.9 ft.
	Tonnage	2.57
Rig:	Gaff mainsail and foresail	

The *Atlanta* was probably the first lobsterboat built for engine propulsion. She is said to have been very similar in design to the old lobsterboats, and in spite of the engine, always carried her mainsail and staysail. She was owned by Philip O'Dwyer of Marsh, Aughadown. At some stage, she was transferred to the Port of Dublin Registry where she was registered under the number D195.

Barque (see ***Star of the Sea* S227**)

Belle-Isle

Builder:	Probably the Pyburns, Toormore
Date built:	Unknown
Rig:	Dipping lugsail

Said to have been a lobsterboat owned by Johnny 'Jer Mick' O'Driscoll of Long Island. As well as lobstering, O'Driscoll also used the *Belle-Isle* for mackerel fishing and trawling under sail. This was probably the boat that was sunk while moored off Long Island pier when a boat from the mainland collided with her. She was later re-floated and repaired. The boat from the mainland was bringing people to the regular Sunday evening session of music and dance on Long Island pier. O'Driscoll also owned the *St Catherine* S30 at one stage.

Bonnie Lass (see ***Star of the Sea* S227**)

Britannia

Owned by Michael O'Driscoll of Heir Island up to the 1890s, she had the usual gaff rig. She was among the boats that competed in the lobsterboat race at Baltimore Regatta in 1896.

Bruiser

Builder:	The Pyburns, Toormore
Date built:	Late 1890s
Rig:	Dipping lugsail initially; gaff mainsail and foresail later

This boat was owned by Johnny O'Sullivan, Long Island. Johnny later bought the Cape Clear mackerel yawl, *St Veronica*, but the O'Sullivan family also continued to fish the *Bruiser* for lobsters. She was hauled up on Long Island in the late 1940s where she gradually deteriorated and rotted away.

Catherine (see ***St Catherine* S30**)

***Catherine of Baltimore* S611**

Builder:	Willie Skinner, Rathmore, Baltimore	
Date built:	1910	
Date registered:	9/6/1910	
Registered dimensions:	Length overall:	25.0 ft.
	Breadth:	6.8 ft.
	Depth:	3.2 ft.
	Tonnage:	3.18
Rig:	Gaff mainsail and foresail	

This boat was owned by Cornelius Cadogan of Poulnacallee, Aughadown. She was inherited by his son Patrick, and was registered in his name in 1924. Her registration was closed in 1937 when she 'ceased to be a fishing boat'. It is said that the Cadogan family of Poulnacallee were known as the 'Kings' because they once had two lobsterboats.

Con Florrie S235

Builder:	Henry Skinner, Baltimore	
Date built:	1933	
Date registered:	11/8/1933	
Registered dimensions:	Length of keel:	24.0 ft.
	Length overall:	27.5 ft.
	Breadth:	7.3 ft.
	Depth:	3.5 ft.
	Tonnage:	3.55
Rig:	Gaff mainsail and foresail	

The *Con Florrie* was one of the first, and one of the very few, lobsterboats built for engine propulsion. She had greater rise of floor midships than the older lobsterboats, but otherwise was very similar in design. She carried a gaff mainsail and staysail as well as her 6/7 h.p. Kelvin Ricardo.

The *Con Florrie* was originally registered in the name of Thomas Ganey, Crookhaven but was fished by John O'Driscoll, also of Crookhaven, and was named after O'Driscoll's uncle and father. Willie O'Mahony of Crookhaven bought her in the early 1950s, and she was registered in his name in 1955. O'Mahony raised her gunnels and built a foredeck on her, and changed her name to *St Peter*. O'Mahony was the last registered owner of her, but in the late 1950s she was bought by John Cross of Baltimore and used as a pleasure boat. Her next owner was Charlie O'Dowd of Kinsale who changed her name to *St Ann*. In the 1960s, she was bought by Eamonn English of Cobh. At this stage, she had a short afterdeck and a bridge-deck over her engine as well as the foredeck, and had also gained a mizen mast. English replaced her engine with a diesel Lister, and had the bridge-deck removed, her cabin extended and a wheel-house built by Jack O'Driscoll, Crosshaven. The English family used her as a pleasure boat down to the early 1980s. In 1997, she was laid up at Cork Dockyard. She was bought by an organisation known as Cork Harbour 2000 which had plans for her restoration.

The *Con Florrie* at Cork Dockyard in 1998.
Photograph: Cormac Levis

Coolmain (see ***Victory II* S14**)

Dodo

Builder:	The Pyburns, Toormore
Date built:	Probably 1880s or 1890s
Rig:	Dipping lugsail

The *Dodo*, nicknamed the *Rolling Dodo*, was originally owned by Paddy 'Denny' O'Driscoll of Westland, Long Island. She was sold to Charlie O'Regan, also of Westland, Long Island. It is said that she was about twenty-four feet in overall length.

Eleanor

Builder:	Richard Pyburn, Heir Island
Date built:	1890s or early 1900s
Rig:	Gaff mainsail and foresail

The *Eleanor* was owned by the Minihanes of Paris, Heir Island. She was mainly fished by Denis Minihane.

Eliza S750

Builder:	Unknown	
Date built:	Unknown	
Date registered:	5/7/1920	
Registered dimensions:	Length of keel:	22.0 ft.
	Tonnage:	3.0
Rig:	Gaff mainsail and foresail	

The *Eliza* was owned by Daniel Dwyer of Collatroum, Aughadown. Her registration was closed on 18/6/1935 when she was 'broken up'.

Ellen

This boat was owned by Charlie O'Regan, West Calf and came second in a race for Second Class sail boats at Schull Regatta in 1885. On that occasion, she was sailed by Charlie's son Dan who later had a boat of the same name built for himself (*see Ellen* S15).

Ellen S15 also S247

Builder:	The Pyburns, Toormore	
Date built:	June 1904	
Date registered:	1/8/1904 (S15)	
Registered dimensions:	Length overall:	24.0 ft.
	Breadth:	6.3 ft.
	Depth:	5.0 ft. (error)
	Tonnage:	2.75
Rig:	Gaff mainsail and foresail	

The *Ellen* was built for Dan O'Regan of West Calf, who was one of the men awarded medals for rescuing Isaac Baker of the *Petrel* in 1892. She was one of the two lobsterboats fished by Dan and his sons (*see Shamrock* S1085). When Dan had the thirty-six-foot mackerel boat *St Dominic* built in 1910, the *Ellen* became known popularly as the *Small Dominic*.

The *Ellen*'s registration was closed on 19/1/1934, but she was re-registered under the number S247 in the name of Cornelius O'Regan of Ardmanagh, Schull on 12/3/1942. In 1944, she was sold to John Cadogan of Cape Clear who fished dogfish with her. By this time, she had a 12/14 h.p. Kelvin engine and a foredeck, and her gunnels had been raised. In 1948, she returned to the lobster fleet when she was bought by Paddy and John O'Regan of Ballycummisk, Ballydehob, who were sons of Rickie O'Regan, West Skeam, one-time owner of the *Lily* S125 and the *Flower of May* S350. In the late 1950s, the *Ellen* was sold to Michael Crowley of Bere Island. Her registration was closed finally when she ceased fishing in 1963.

Ellen S969

Builder:	Unknown	
Date built:	Unknown	
Date registered:	5/7/1927	
Registered dimensions:	Length of keel:	22.0 ft.
	Tonnage:	2.5
Rig:	Gaff mainsail and foresail	

This boat was registered in the name of Michael Cadogan of Heir Island. Her registration was closed on 21/2/1934 when it was stated that she 'ceased to be a fishing boat'.

Fear Not

Builder:	Richard Pyburn, Heir Island
Date built:	1890s or early 1900s
Rig:	Gaff mainsail and foresail

The *Fear Not*, said to have been about twenty-eight feet in length, was originally owned by Mike Cahalane of The Reen, Heir Island. She was a good sailer in strong wind. Cahalane also fished mackerel for some years with his ten-ton dandy, *St Nicholas*. In the 1940s, Timothy John Cadogan of Ardagh, Aughadown used the *Fear Not* as a sandboat on the Ilen River. It is said that Cadogan bought her from Cahalane for five pounds.

Flower of May S350

Builder:	Richard Pyburn, Heir Island	
Date built:	Unknown	
Date registered:	22/3/1901	
Registered dimensions:	Length overall:	24.0 ft.
	Tonnage:	2.0
Rig:	Gaff mainsail and foresail; a jib was set for regattas	

The *Flower of May* was one of two lobsterboats owned and fished by Rickie O'Regan and his family of the West Skeam Island. She was more commonly known on the Skeam Islands as the *Green Yawl*. She was bought from O'Regan in the early 1930s by Timmy O'Driscoll of Schull, who was originally from West Calf Island. When O'Driscoll owned her, she was known by some people as the *Mayflower*. O'Driscoll did some pollock fishing with her and raced her very successfully at Schull and Baltimore regattas. It is said that she was wrecked by a storm one Christmas night while moored at the Mill Cove in Schull Harbour.

Gold Seeker

Builder:	The Pyburns, Toormore
Date built:	Unknown
Rig:	Dipping lugsail

Little is known of the early history of this boat. In the 1930s and the 1940s, she was in a laid-up state on Long Island where she gradually decomposed. Her owner, Con 'Hayes' O'Driscoll, also had the mackerel yawl, *St Ita*. He also did some lobstering in a small lugsail yawl about seventeen feet in length. The *Gold Seeker* was one of the boats referred to in Long Island, as a longboat or six-oar yawl.

Green Yawl (see ***Flower of May* S350**)

Hannah

This boat, owned by Cornelius Cahalane, The Reen, Heir Island came second in the lobsterboat race at Baltimore Regatta in 1924. She had the usual gaff rig of the Heir Island lobsterboats.

Hanorah S463

Builder:	Richard Pyburn, Heir Island	
Date built:	1892 or early 1893	
Date registered:	24/6/1893	
Registered dimensions:	Length overall:	24.6 ft.
	Breadth:	6.6 ft.
	Depth:	3.0 ft.
	Tonnage:	5.0 (miscalculation)
Rig:	Gaff mainsail and foresail	

The *Hanorah* was built for Con Harte, The Reen, Heir Island. Informants differ as to what she cost, some saying fourteen pounds and some sixteen. She was among the boats that competed in the lobsterboat race at Baltimore Regatta in 1896. She was inherited by Con's son Dan who fished her until about 1951. In the 1950s, Dan's sons, Neily and Seán, bought boats of their own; and in 1958, the *Hanorah*, which had been laid up since Dan ceased fishing her, was sold to Seán Barnett of Schull.

Barnett employed Denny 'Kilderry' Keohane, a well-known shipwright from Ballylickey, Bantry, to raise her gunnels and fit her out with a foredeck. The *Hanorah*, which had been well maintained over the years, was found to be in very good condition and planked with pitch pine on oak frames. She was fitted with a new Coventry Victor 10 h.p. diesel engine and a pot-hauler. It is said that she was the first boat in the locality to have the engine, pot-hauler and a wheel all positioned up forward so that one man could control the boat and haul the pots simultaneously, thus reducing the number of crew from three to two.

Barnett employed Den Murphy from Long Island and Pat Burke, originally from Heir Island, to fish the *Hanorah* locally in Roaringwater Bay. Den Murphy, a die-hard traditionalist, was very reluctant to use the hauler initially, and continued to use the old-style withy pots. In spite of the new diesel engine, which had replaced the *Hanorah*'s gaff rig, he always carried a lugsail on a new mast footed on the foredeck and a pair of oars.

In 1959, Den Murphy bought the *Hanorah* from Seán Barnett. In the late 1960s, he had some repairs carried out by Paddy

Hegarty of Oldcourt. Murphy continued to fish her right down through the 1970s, and was probably the last lobsterman to make and fish the old withie pots.

The *Hanorah* was not to escape without her share of tragedy, however. After many years of bringing her crew safely through storms and high seas along the south coast, Michael O'Sullivan of Long Island, Den Murphy's crewman, was drowned out of her in Roaringwater Bay.

In 1980, Den Murphy sold *Hanorah* to John O'Callaghan of South Schull who had a new foredeck fitted by Paddy Hegarty, and used her for pleasure fishing. About 1984, she was sold to a man by the name of O'Donnell from Co. Clare. The *Hanorah*, however, had no intention of leaving Roaringwater Bay; on two successive occasions the trailer on which she was about to be transported broke down before ever leaving Schull. O'Donnell eventually removed her engine and abandoned her at the Mill Cove in Schull Harbour where she gradually deteriorated, for a period of about fifteen years, until she was finally rescued by Nigel Towse of Sherkin Island in 1999. With much of her original structure still intact, Nigel is currently in the process of restoring her fully.

Irish Leader S411

Builder:	Flor O'Neill, Baltimore	
Date built:	1909	
Date registered:	6/7/1909	
Registered dimensions:	Length overall:	27.0 ft.
	Breadth:	6.8 ft.
	Depth:	3.4 ft.
	Tonnage:	3.52
Rig:	Gaff mainsail and foresail	

This boat, better known as simply the *Leader*, was registered in the name of Denis Minihane, Midland, Heir Island but was always fished by Denis' brother Con. She had the reputation of being a good sailer. In 1948, Con Minihane bought the *Safe Return* S198 and sold the *Irish Leader* to a man by the name of Motherway from Ballycotton. She was later sold to someone in Crosshaven. Her name was possibly inspired by a large Arklow-based mackerel drifter, also called the *Irish Leader*, that would have fished out of Baltimore in the 1890s.

Jollyboat

Builder:	The Pyburns, Toormore
Date built:	*c.* 1920
Rig:	Dipping lugsail

The *Jollyboat* was owned by Johnny Goggin of Eastland, Long Island. On Long Island, the name of the boat was sometimes compounded with the word 'boat'.

Karen Marie (see ***Two Fishes* S722**)

Láirín

Builder:	The Pyburns, Toormore
Date built:	Unknown
Rig:	Dipping lugsail

The *Láirín* (meaning 'little mare') was owned by Florrie 'Batt' Whooley of Eastland, Long Island. A small boat, about twenty feet in length, she was inherited by his son Paddy. Florrie 'Batt' also owned the *Mayflower* S752.

Laveneer

Builder:	Thought to be Willie Skinner, Rathmore, Baltimore
Date built:	1890s or early 1900s
Rig:	Gaff mainsail and foresail

This boat was owned by Michael (Mikeen) O'Neill, Heir Island.

Leadboat (see ***Shamrock* S1085**)

Leader (see ***Irish Leader* S411**)

Lily S125

Builder:	Richard Pyburn, Heir Island	
Date built:	1905	
Date registered:	19/7/1905	
Registered dimensions:	Length of keel:	25.0 ft.
	Length overall:	26.5 ft.
	Breadth:	6.8 ft.
	Depth:	3.4 ft.
	Tonnage:	3.77
Rig:	Gaff mainsail, foresail and jib	

The *Lily* was one of two lobsterboats owned and fished by Rickie O'Regan and his sons of the West Skeam Island. She was popularly known as the *White Yawl* on the Skeam Islands. It is said that she was sold to Patsy Murphy of Heir Island in the 1930s, but his name does not appear in the *Register*. In 1941, she was bought by James Nolan of Baltimore, who sold her to Denis Shanahan of Glandore in 1942. In 1948, she was bought by John Kelleher, Glandore and he sold her to Michael Kelleher, Glandore the following year.

Margaret (see *St Margaret* S47)

Marguerite

This boat won the race for lobsterboats at Crookhaven Regatta in 1907. According to the newspaper report on the regatta, she was owned by John Murphy. This was probably John Murphy of Long Island, who was originally from Heir Island. He had the *St Patrick* S368 built in 1909, and his son Den owned the *Hanorah* in the 1960s and 1970s.

Maria S9O9

Builder:	Unknown, but built at Baltimore	
Date built:	1912	
Date registered:	18/10/1926	
Registered dimensions:	Length of keel:	22.6 ft.
	Length overall:	27.6 ft.
	Breadth:	6.11 ft.*
	Depth:	1.6 ft.‡
	Tonnage:	3.47
Rig:	Gaff mainsail and foresail	

* *This seems to indicate that the dimensions of this boat were actually given in feet and inches rather than feet and tenths of a foot.*
‡ *The depth of the boat seems to have been underestimated.*

This boat was owned by John Fitzgerald of Heir Island. The last entry in the registration is dated 1/6/1942 and she was still registered in his name. The registration is marked 'change of ownership'. It is probable that she was also registered before this in the previous owner's name.

Mary S3

Builder:	The Pyburns, Toormore	
Date built:	12/6/1903	
Date registered:	30/7/1904	
Registered dimensions:	Length overall:	24.5 ft.
	Breadth:	6.8 ft.
	Depth:	3.2 ft.
	Tonnage:	3.58
Rig:	Gaff mainsail and foresail	

The *Mary* was built for Patrick O'Regan of Schull. This was the Patrick O'Regan who was honoured for his part in saving the life of Isaac Baker of the *Petrel* in 1892, when he was living on West Calf Island. O'Regan fished 'down east', as well as 'up north', which was unusual for lobstermen from the west side of the bay. He used the *Mary* for line fishing and trawling as well as lobstering; and although her registered dimensions fall within the range associated with lobsterboats, she is remembered as a slightly more substantial craft than the typical lobsterboat. She was fitted with a foredeck at an early stage, probably the early 1920s, which is unusual for a lobsterboat. Her gunnels were raised at some stage, and a Bolinder engine was installed in her in the 1940s. In 1956, she was bought by a partnership of three men from Cork, one of them named O'Mahony, who based her at Kinsale (*see photograph* p. 95).

Maryboat

Builder:	The Pyburns, Toormore
Date built:	Unknown
Rig:	Dipping lugsail

The *Maryboat*, owned by Mike 'Seán Dónaill' O'Regan of Eastland, Long Island, was one of the boats referred to as a longboat or six-oar yawl. Even after the advent of motorised boats, this boat, manned by six oarsmen and a cox, was the boat preferred for making emergency trips to the mainland in bad weather. Six good oarsmen and an experienced cox could manoeuvre a boat better when coming alongside a landing place in a heavy swell and strong winds, than a man at the helm of a motor boat. The *Maryboat* was the boat in which Peter O'Driscoll of Long Island was killed by lightning in the early 1920s (*see photograph* p. 16).

Mary Ellen (see ***Two Fishes* S722**)

Mary Ellen S1016

Builder:	Unknown, but built at Baltimore	
Date built:	1903	
Date registered:	10/7/1903	
Registered dimensions:	Length overall:	26.2 ft.
	Breadth:	7.1 ft.
	Depth:	3.1 ft.
	Tonnage:	3.79
Rig:	Gaff mainsail and foresail	

This boat was owned by John Minihane, Heir Island. Her registration was closed in 1925 when she was 'sold out of port'.

Mary Ellen S1013

Builder:	Flor O'Neill, Baltimore	
Date built:	1903	
Date registered:	19/7/1903	
Registered dimensions:	Length of keel:	22.0 ft.
	Length overall:	26.9 ft.
	Breadth:	6.9 ft.
	Depth:	3.6 ft.
	Tonnage:	3.81
Rig:	Gaff mainsail, foresail and jib; a topsail was also set for regattas	

The *Mary Ellen* was built for Denis Minihane Mór of Cunnamore in the parish of Aughadown. She passed on to his son, also Denis, and grandson, Neily. She had an excellent record at regattas over many years, especially at Kinsale and, in later years, Baltimore. Neily Minihane installed an engine in her in 1945 and sold her some years later. After this, she was primarily used as a pleasure boat. She was based at Crosshaven for some years in the early 1960s. She also spent some time at Kinsale and on the River Shannon but seems to have eventually ended up on the English canals. She was last spotted in 1978 at Leamington, Warwickshire in the heart of England by Neily's brother Denis.

Mary Hannah S896

Builder:	Richard Pyburn, Heir Island	
Date built:	Unknown	
Date registered:	2/7/1896	
Registered dimensions:	Length overall:	24.0 ft.
	Breadth:	6.6 ft.
	Depth:	3.0 ft.
	Tonnage:	2.0
Rig:	Gaff mainsail and foresail	

The *Mary Hannah* was originally owned by Michael McCarthy of Paris, Heir Island. She later passed on to Michael's son Jeremiah. She was broken up in 1960. Jeremiah's son Timmy had bought the *Safe Return* S198 in 1954. (*See photograph* p. 12.)

***Mary Hannah* S1055** also **S258**

Builder:	Unknown	
Date built:	Unknown	
Date registered:	11/6/1898	
Registered dimensions:	Length of keel:	22.0 ft.
	Length overall:	22.0 ft. (apparent error)
	Breadth:	6.5 ft.
	Depth:	3.0 ft.
	Tonnage:	2.5
Rig:	Gaff mainsail, foresail and jib	

This boat was originally registered in the name of Michael O'Neill, Heir Island. The registration was closed on 10/11/1941 when, it is stated, she 'ceased to be a fishing vessel'. She was re-registered on 13/6/1942 in the name of John O'Donovan, Heir Island under the number S258. It is stated that she had been rebuilt that year. Her registered dimensions correspond with those of the first registration, including the apparent error in her overall length. Presumably, this was copied directly from the original registration by the official making the entry. Her rig also remained unchanged, but an auxiliary motor had been fitted and her tonnage was reduced to 1.93. Her last registered owner was Patrick S. Dempsey of The Cove, Baltimore. That entry was made on 3/7/1942.

***Mary Kate* S391**

Builder:	Flor O'Neill, Baltimore	
Date built:	1892	
Date registered:	29/6/1892	
Registered dimensions:	Length of keel:	22.0 ft.
	Length overall:	27.0 ft.
	Breadth:	7.0 ft.
	Depth:	4.0 ft.
	Tonnage:	4.0
Rig:	Gaff mainsail and foresail	

The *Mary Kate* was built for Michael McCarthy of Paris, Heir Island and inherited by his son Jim. Around 1940, Jim had major repairs carried out on her by Willie Skinner of Rathmore, Baltimore at the cost of twenty-five pounds. In 1951, she was bought by Timothy Daly of Union Hall, who sold her to Joseph Moloney of Rineen, Castlehaven in 1958. Joe and his brother John carried out some further restoration at this time and installed an Austin car engine. They fished lobsters and herring with her for a short period before she was badly damaged when she dragged her anchor during a gale at Union Hall. Joe repaired her once more, but she remained unused for some years before she was eventually broken up and burned.

Mayflower S752

Builder:	The Pyburns, Toormore	
Date built:	1901	
Date registered:	23/10/1908	
Registered dimensions:	Length overall:	26.0 ft.
	Breadth:	6.9 ft.
	Depth :	3.3 ft.
	Tonnage:	4.2
Rig:	Unable to ascertain whether dipping lug or gaff rigged	

The *Mayflower* was owned by Florrie 'Batt' Whooley, Long Island, who also owned the *Láirín*. She was broken up in 1925.

Mayflower (see *Flower of May* S350)

Mermaid

Builder:	The Pyburns, Toormore
Date built:	Unknown
Rig:	Dipping lugsail

This boat was owned by Florrie 'Hayes' O'Driscoll of Westland, Long Island. In the early 1940s, O'Driscoll had a mackerel boat built which was known as the *St Ann*.

Nonpareil S1022

Builder:	Flor O'Neill, Baltimore	
Date built:	1903	
Date registered:	10/7/1903	
Registered dimensions:	Length of keel:	27.0 ft.
	Length overall:	27.8 ft.
	Breadth:	7.1 ft.
	Depth:	3.3 ft.
	Tonnage:	3.86
Rig:	Gaff mainsail and foresail	

The *Nonpareil*, possibly named after an English steamer of that name, was built for Michael J. O'Driscoll ('Mikey Merchant') of Heir Island. She passed on to his son, Pat O'Driscoll, and was registered in his name in 1934. In 1941, she was sold to John Beamish of Sherkin Island, who employed Simmie McCarthy, a Sherkin Island shipwright, to raise her gunnels and fit a foredeck. An engine was also installed. Beamish used her mainly for long-lining for dogfish and for mackerel fishing. Around 1950, she was sold to Denis Nolan, Sherkin Island who fished lobsters with her. In 1960, she was registered in the name of James Nolan, Sherkin Island, with Denis Nolan named as her skipper. Some years later, she caught fire off Fordree Cuas on the south side of Sherkin Island. Denis Nolan escaped by swimming ashore at Carrigsle. The *Nonpareil* was taken in tow by Baltimore lifeboat but foundered outside of Horseshoe Harbour, Sherkin Island.

Oileán Garbh (see ***Starr* S694**)

Othello S459

Builder:	Flor O'Neill, Baltimore	
Date built:	1909	
Date registered:	5/10/1925	
Registered dimensions:	Length of keel:	22.5 ft.
	Length overall:	27.4 ft
	Breadth:	7.8 ft.
	Depth:	3.4 ft.
	Tonnage:	4.93
Rig:	Gaff mainsail and foresail	

This boat was also known as the *Sage*.

One morning in the mid-1950s in the port of Padstow, Cornwall, a bewildered John Burke stood on the deck of a large trawler as she headed out of the harbour. As the trawler was pulling away from the pier, Burke had caught sight of a small boat moored among the large, modern fishing boats. Although he had sight of her for only a few minutes, there was no doubt about it, the last time he had seen a boat like this he had been back home on Heir Island. Not only that, but he felt sure that this was his brother's boat, the *Sage*. But what was she doing in Padstow? Was his imagination playing tricks on him? Had his brother Pat navigated all the way from Roaringwater Bay to Padstow in his lobsterboat? John Burke, heading off on a long fishing trip, had nine perplexing days to ponder these questions before he was back in port and able to get the answers. In actual fact, the *Sage* had been bought from Pat Burke by a man from Cornwall who, with only his cat and his dog for crew, had cruised her from Heir Island to Castletownbere and back to Sherkin Island before setting out for Padstow.

The *Sage* was originally part of the Aughadown lobster fleet, and was built for someone who lived on the banks of the Ilen River. She was later owned by Jimmy McCarthy of Sherkin Island, who sold her to Denny 'Thade' O'Driscoll of Baltimore in 1925. O'Driscoll raised her gunnels, fitted a foredeck and installed an engine. During the years of World War II, she was used for fishing dogfish. In 1950, she rejoined the lobster fleet when she was bought by Pat Burke. According to her registration records, she was bought by Jeremiah McSweeney of Baltimore in 1954 but this sale, apparently, never actually took place. Instead, Burke sold her to a Cornish man; and as we have seen, the *Sage* ended up in Padstow.

Pride S124

Builder:	Unknown, but built at Baltimore	
Date built:	1911	
Date registered:	11/6/1931	
Registered dimensions:	Length of keel:	20.0 ft.
	Length overall:	22.0 ft.
	Breadth:	6.3 ft.
	Depth:	2.5 ft.
	Tonnage:	1.56

The *Pride* belonged to Jeremiah McCarthy of Heir Island. Her registration was closed in 1958 when she was broken up.

Pride of Toe Head S894

Builder:	Henry Skinner, Baltimore	
Date built:	1912	
Date registered:	20/5/1912	
Registered dimensions:	Length of keel:	23.0 ft.
	Length overall:	28.0 ft.
	Breadth:	7.4 ft.
	Depth:	3.3 ft.
	Tonnage:	4.77
Rig:	Gaff mainsail and foresail; a jib and topsail were set for regattas	

The *Pride of Toe Head* was originally built for Laurence Daly of Lickowen, Castletownsend who apparently did not put her to much use and soon sold her to Patrick Whooley of Lisaree, Aughadown. Whooley sold her on to a partnership of three men in Sherkin Island. These men, one of whom was named O'Donovan, used her for 'fall fishing', i.e. mackerel fishing during the autumn season.

In 1917, five years after she was built, she was bought by Timothy O'Donovan of Heir Island for nine pounds. The *Pride of Toe Head* had a foredeck at that stage which O'Donovan removed so that she could be rowed up forward for lobster fishing. He had other minor alterations carried out by the O'Driscolls of The Cove, Baltimore, including the fitting of the usual locker in the bow. She was said to be a very good sailer in strong wind.

The *Pride of Toe Head* was inherited by Timothy's son Neily who installed a 13/15 h.p. petrol-paraffin Kelvin, raised her gunnels and fitted a foredeck in the early 1950s. Neily fished her down to the mid-1960s.

In 1979, the *Pride of Toe Head* was sold to John O'Leary of Schull. She was broken up in 1983.

Rolling Dodo (see ***Dodo***)

Safe Return

Builder:	Flor O'Neill, Baltimore
Date built:	Unknown
Rig:	Gaff mainsail, foresail, jib and topsail

The *Safe Return* was built for Jim Barry of East Skeam Island probably towards the end of the nineteenth century. She was reputed to be a very good sailer (see the description of her on p. 16). She was eventually sold to Dick Salter of Baltimore who was responsible for having the usual alterations carried out to her sheer line, and for having a foredeck and engine installed. Salter did some mackerel fishing with her. Her next owner was John Hosford, Skibbereen who sold her to someone in the Kinsale area.

***Safe Return* S12** also **S141**

Builder:	Richard Pyburn, Heir Island	
Date built:	1898	
Date registered:	17/4/1929 (S12) and 22/9/1940 (S141)	
Registered dimensions:	Length of keel:	21.0 ft.
	Length overall:	26.0 ft.
	Breadth:	6.5 ft.
	Depth:	3.5 ft.
	Tonnage:	2.66
Rig:	Gaff mainsail and foresail	

This boat was registered in the name of John Dwyer, Collatrum, Aughadown in 1929. In 1935, that registration was closed when the boat was stated to have been 'broken up'. However, she reappears on the *Register* in 1940 when she was registered under the number S141 in the name of John Hurley, Union Hall. At this stage, she had an engine. She was once again said to have been 'broken up' when her registration was finally closed in 1950.

Safe Return S198

Builder:	Henry Skinner, Baltimore	
Date built:	1916	
Date registered:	28/6/1941	
Registered dimensions:	Length of keel:	22.0 ft.
	Length overall:	27.0 ft.
	Breadth:	7.7 ft.
	Depth:	4.2 ft.
	Tonnage:	3.93
Rig:	Gaff mainsail, foresail and jib	

This boat was built for Joe O'Driscoll, Schull, formerly of East Calf Island. O'Driscoll used her primarily for transporting livestock back and forth between the mainland and his farm on the island. She was fitted with an engine at a very early stage, and had a removable section in her top two planks to facilitate the loading and unloading of cattle. In 1943, she was bought by Jeremiah Barnett of Lowertown, Schull who used her for trawling. In 1946, she was bought by Paddy O'Sullivan, Colla, a son of the owner of the famous mackerel yawl, *Shamrock*. It was probably around this time that she was fitted with a foredeck. O'Sullivan sold her the

The *Safe Return* at St Mullins on the River Barrow in Co. Carlow, 2001.
Photograph: Norcott Roberts

following year to Jack O'Brien of Ballycummisk, Ballydehob who used her as a sandboat and, for a brief period, as a ferry for schoolchildren from Horse Island attending school at Rossbrin.

In 1948, the *Safe Return* joined the lobster fleet when she was sold to Con Minihane of Heir Island. Minihane was already the owner of the *Irish Leader* S411, which he sold the following year. In 1954, the *Safe Return* was sold to Timmy McCarthy, Paris, Heir Island whose father owned the *Mary Hannah*. The *Safe Return* remained with the McCarthy family for many years and narrowly escaped being wrecked on Coosheen Point when Timmy's brother John was returning at night from Schull having sold lobsters there. Her registration was closed in 1978 when, it was stated, she 'ceased to be a fishingboat'. At this time, she was, in fact, sold to Frank Bohane, Garrettstown who sold her on to a Mr O'Neill in Cork, who in turn sold her on to Derek Kelly in Bandon. Kelly based her at Courtmacsherry for some years, and had her repaired by John Maloney of Union Hall. In the mid-1990s, the *Safe Return* took to the inland waterways when she was sold to Rob Robinson who based her at St Mullins on the River Barrow in Co. Carlow. In 2001, she was photographed there by the ever-vigilant, classic-boat enthusiast, Norcott Roberts, who recognised her West Cork build.

Sage (see ***Othello* S459**)

St Ann (see ***Con Florrie* S235**)

St Anna S136

Builder:	Christopher and Richard Pyburn, Heir Island	
Date built:	Unknown	
Date registered:	13/6/1887	
Registered dimensions:	Length of keel:	24.0 ft.
	Length overall:	25.5 ft.
	Breadth:	7.5 ft.
	Depth:	3.5 ft.
	Tonnage:	3.5
Rig:	Gaff mainsail, foresail and jib	

The *St Anna* was initially registered in the name of Patrick O'Neill of Heir Island, and her skipper was given as Dan O'Neill. She left the fleet in 1939 when she was sold to Cornelius Hurley of Union Hall, and in 1942 to Davis Wolfe of Tragumna. In 1947, she returned to

Union Hall where she was owned by John Nolan. An engine had been fitted at this stage. Her registration states that she was transferred to Youghal on 21/6/1948.

St Catherine S30

Builder:	Jack Pyburn, Crookhaven	
Date built:	1906	
Date registered:	21/8/1936	
Registered dimensions:	Length of keel:	22.8 ft.
	Length overall:	27.9 ft.
	Breadth:	7.9 ft.
	Depth:	4.0 ft.
	Tonnage:	3.97
Rig:	Gaff mainsail and foresail	

This boat originally belonged to Den 'Paddy' O'Sullivan of Colla, Schull, and was originally called the *Catherine*, after his wife, rather than *St Catherine*. She was mainly fished by Den's brother Mick, who would have used her primarily for mackerel and herring fishing. In 1910, Mick had his famous mackerel yawl *Shamrock* built; and sometime after this, possibly in the 1920s, the *Catherine* was sold to Johnny 'Jer Mick' O'Driscoll of Midland, Long Island for twelve pounds. It was Johnny 'Jer Mick' who fished lobsters with her. He also did a lot of mackerel fishing. In 1936, she was sold to Richie Daly of Schull, and in 1942, she passed on to his brother Willie, of Toormore. The Dalys used her mainly for pollock fishing and drawing gravel. She had an engine at this stage. Her registration was closed in 1948.

St Helena S19

Builder:	Unknown	
Date built:	Unknown	
Date registered:	2/2/1899	
Registered dimensions:	Length of keel:	24.0 ft.
	Length overall:	27.0 ft.
	Breadth:	6.4 ft.
	Depth:	2.9 ft.
	Tonnage:	3.37
Rig:	Gaff mainsail and foresail	

The *St Helena* was originally owned by Michael O'Neill, Heir Island. In 1941, she was sold to Denis Daly of Cape Clear.

St Margaret S47

Builder:	Willie Skinner, Rathmore, Baltimore	
Date built:	1925	
Date registered:	7/5/1925	
Registered dimensions:	Length of keel:	23.0 ft.
	Length overall:	27.2 ft.
	Breadth:	7.5 ft.
	Depth:	3.4 ft.
	Tonnage:	4.1
Rig:	Gaff mainsail and foresail	

This boat, owned by Johnny Harte of The Reen, Heir Island, was probably the last truly traditional sailing lobsterboat to be built. In 1949, she was sold to Timothy Keohane, Rineen, Union Hall. In 1959, her registration was closed when she was deemed to be unseaworthy.

St Patrick S700

Builder:	The Pyburns, Toormore	
Date built:	1907	
Date registered:	25/6/1908	
Registered dimensions:	Length overall:	25.7 ft.
	Breadth:	6.6 ft.
	Depth:	2.2 ft.
	Tonnage:	2.97
Rig:	Gaff mainsail and foresail	

This boat was registered in the name of Patrick O'Regan, Long Island, but was fished by Mikey 'Mick Óg' O'Regan. This was the boat with which Mikey 'Mick Óg' recovered the lifeboat and bodies from the *Lusitania* disaster. Mikey 'Mick Óg' later bought the Cape Clear mackerel yawl, *St Brendan*. According to the *Register*, *St Patrick* was 'sold out of port' in 1929.

St Patrick S368

Builder:	The Pyburns, Toormore	
Date built:	1909	
Date registered:	15/6/1909	
Registered dimensions:	Length overall:	26.8 ft.
	Breadth:	7.7 ft.
	Depth:	3.1 ft.
	Tonnage:	4.57
Rig:	Gaff mainsail and foresail	

This boat was owned by Jack Murphy of Midland, Long Island, who was originally from Heir Island. Jack's son Den owned the *Hanorah* in the 1960s and 1970s.

St Patrick S567

Builder:	The Pyburns, Toormore	
Date built:	1909	
Date registered:	28/1/1910	
Registered dimensions:	Length overall:	28.0 ft.
	Breadth:	7.9 ft.
	Depth:	3.3 ft.
	Tonnage:	5.07
Rig:	Gaff mainsail and foresail	

This boat was owned by Denny 'Donnchadh' O'Donovan of Long Island, and later by his son, 'Big Pat'. In 1914, Pat had the mackerel yawl *St Finbarr* built. In 1926, the *St Patrick*'s registration was closed when she 'ceased to be a fishing boat' (*see photograph* p. 70).

St Patrick and Mary S249

Builder:	Unknown	
Date built:	Unknown	
Date registered:	6/6/1921	
Registered dimensions:	Length of keel:	22.0 ft.
	Length overall:	26.3 ft.
	Breadth:	7.0 ft.
	Depth:	3.0 ft.
	Tonnage:	4.29
Rig:	Gaff mainsail and foresail	

This boat was owned by John Harte, of Laheratanavally, Aughadown and originally of Heir Island. This was the illustrious John

Harte who survived the 1904 drowning tragedy and played major parts in the *Christian Wilhelm* and *Thomas Joseph* rescues, among others. The *St Patrick and Mary* was inherited by his son, also named John, who fished her for many years before she was finally hauled up on the shore at Laheratanavally around 1960, where she gradually decayed.

St Peter (see ***Con Florrie* S235**)

***Shamrock* S1085**

Builder:	Unknown	
Date built:	Unknown	
Date registered:	20/9/1898	
Registered dimensions:	Length of keel:	22.0 ft.
	Tonnage:	2.0
Rig:	Dipping lugsail	

This boat was owned by Dan O'Regan, West Calf Island, and is almost definitely the boat that was better known as the *Leadboat*. She was one of two lobsterboats fished by Dan and his family (*see Ellen* S15). Her registration was closed on 31/12/1928 when she was reported 'lost'. Dan O'Regan had been drowned out of her when going ashore on the West Calf around this time.

***Shamrock* S1100**

Builder:	Unknown, but built at Baltimore	
Date built:	1909	
Date registered:	6/8/1913	
Registered dimensions:	Length overall:	27.0 ft.
	Breadth:	6.8 ft.
	Depth:	3.4 ft.
	Tonnage:	4.12
Rig:	Gaff mainsail and foresail	

This boat was owned by Phil Dwyer and John O'Mahony of Lake Marsh, Aughadown. The last entry in her registration is dated 13/5/1925.

Skellig Bell **S116**

Builder:	Unknown, but built at Baltimore	
Date built:	1920	
Date registered:	22/8/1940	
Registered dimensions:	Length overall:	20.0 ft.
	Breadth:	5.5 ft.
	Depth:	3.0 ft.
	Tonnage:	1.49
Rig:	Dipping lugsail	

This boat was owned by William Griffin of Long Island. She was broken up in 1951.

Small Dominic (see ***Ellen*** **S15**)

Star of the Sea **S227**

Builder:	Jack Pyburn, Crookhaven	
Date built:	1905	
Date registered:	25/5/1906	
Registered dimensions:	Length of keel:	23.6 ft.
	Length overall:	26.7 ft.
	Breadth:	7.0 ft.
	Depth:	3.7 ft.
	Tonnage:	3.61
Rig:	Gaff mainsail, foresail, jib and topsail	

This boat was known by three different names. Needless to say, this multiplicity of names caused some initial difficulties in researching the history of the boat, especially as most informants were not aware that she had more than one name. The issue was further complicated by the fact that two of the names were also the names of other boats. Her registered name, *Star of the Sea*, was also the name of a large, locally owned mackerel drifter as well as the name of one of the French lobster cutters based at Crookhaven. Another of the names by which she was known, especially by the people of Dunmanus Bay, was *Bonnie Lass*, and this was also the name of one of the English lobster cutters. Her owner raced her under the name *Bonnie Lass* at Baltimore Regatta in 1935, but the name by which she was most popularly known was the *Barque*.

The *Star of the Sea* was originally owned by Thomas Scully of the Middle Calf Island. Described by some as the 'pride of the fleet', she is said to have been an excellent sailer. After Thomas Scully moved to Kielbrinogue, Ballydehob, he sometimes put the *Star of the Sea* to work as a sandboat, but lobstering continued to be his main occupation. In 1934, the boat passed on to Thomas' nephew, Cornelius O'Brien of Kielbrinogue, but Thomas continued to fish on her for many years. In the late 1940s, O'Brien raised her gunnels and installed an engine and foredeck. O'Brien continued to fish lobsters and also trawled for shrimps, dredged scallops and drew sand with her. In the 1960s, she was sold to Donal Sheehy, Baltimore who fished salmon with her.

Starr S694

Builder:	The Pyburns, Toormore	
Date built:	1900	
Date registered:	15/6/1908	
Registered dimensions:	Length overall:	26.3 ft.
	Breadth:	7.0 ft.
	Depth:	3.2 ft.
	Tonnage:	4.07
Rig:	Unable to ascertain whether dipping lug or gaff rigged	

This boat, better known as the *Oileán Garbh*, was owned by Connie 'Mick Óg' O'Regan. Oileán Garbh, or Garbh Oileán (Rough Island), is a tiny rocky island off the west end of Long Island. At the end of the lobster season every year, Connie 'Mick Óg' would fit out the *Oileán Garbh* with portable side decks and go 'fall fishing' for mackerel. In 1909, he had the mackerel yawl *Erin* built but also kept possession of the *Oileán Garbh* down to the mid-1920s, at least.

Swan S121

Builder:	Yet to be ascertained *(see below)*	
Date built:	1905	
Date registered:	14/7/1905	
Registered dimensions:	Length of keel:	24.0 ft.
	Length overall:	27.8 ft.
	Breadth:	7.4 ft.
	Depth:	3.6 ft.
	Tonnage:	4.57
Rig:	Gaff mainsail, foresail and jib	

It has yet to be determined who built this boat. There is a tradition that says she was the last boat built by Richard Pyburn of Heir Island. It has also been claimed that she was built by Jack Pyburn of Crookhaven. The *Register*, however, indicates that she was built at Baltimore. The *Swan* was originally owned by Richard Cotter of

Heir Island. By 1935, she was in the possession of John O'Donoghue of Cape Clear who came first with her in the lobsterboat race at Schull Regatta that year. O'Donoghue was Cotter's son-in-law and the owner of a mackerel yawl called the *St Patrick*, but he also did some lobstering with the *Swan*. The *Swan* later underwent the usual modifications of having her gunnels raised and a foredeck and engine fitted, and was used for long-lining for dogfish. In 1941, she was bought by Pat Burke of Heir Island who sold her the following year to John O'Sullivan of Valentia Island, Co. Kerry. In 1949, she was sold to Charles O'Connell, also of Valentia. Her registration was closed on 14/10/1952 when she was reported lost.

TBC S891

Builder:	The Pyburns, Toormore
Date built:	Unknown
Date registered:	12/6/1896
Rig:	Gaff mainsail and foresail

The *TBC* was owned by Thade Scully of Middle Calf Island. Suggestions as to what the initials *TBC* stand for include the most frequent one, 'Thady's Best Clipper'. Accounts of her sailing performance vary. Some claim that she was an 'able' boat, but not very fast. It is also claimed that she sometimes raced the French cutters at regattas and would beat them in light wind. The French cutters were considered very good sailers.

In the 1940s, the *TBC* left the lobster fleet when she was bought by Mikey O'Driscoll of Sherkin Island who carried out the usual procedure of raising her gunnels and fitting a foredeck and engine. O'Driscoll used her for herring fishing and for long-lining for dogfish. In 1950, she was registered in the name of Mikey's son, Pat. She eventually passed on to another son, Finín. One night, the *TBC* set out from Sherkin Island carrying men to join the crew of Baltimore lifeboat, which was setting out in search of a boat that had failed to return to port. The lifeboat collided with the *TBC* in Baltimore Harbour, and sank her. Shortly after, she was refloated and repaired. Eventually, in the late 1960s, she was laid up on a small island opposite Barrack's Quay, on the estuary of the Ilen River, where she ended her days.

Two Fishes S722

Builder:	Richard Pyburn, Heir Island	
Date built:	Unknown	
Date registered:	27/5/1895	
Registered dimensions:	Length of keel:	22.9 ft.
	Length overall:	26.0 ft.
	Breadth:	6.6 ft.
	Depth:	3.0 ft.
	Tonnage:	3.0
Rig:	Gaff mainsail and foresail	

A note in the *Register* indicates that the name of this boat was changed from *Mary Ellen* to *Two Fishes* on 6/6/1941.

The boat was originally owned by Michael O'Neill of Heir Island, and was inherited by his son Danny. She left the fleet in 1941 when she was bought by John J. O'Reilly of Sherkin Island. An engine was installed sometime before she was sold again in 1945. It was probably O'Reilly, also, who raised her gunnels and built a fore-deck. The next known owner of her was Cornelius Murray of Kinsale who used to moor her at the old Lobster Quay in Kinsale Harbour. At this stage, she still carried sails and was sailed up to Passage West when she was bought by Bill Allison. Allison moored her at the Sand Quay in Passage West and removed her engine. She seems to have deteriorated drastically by the time Tony Welland of Monkstown bought her for five pounds around 1970. Welland had the *Two Fishes* completely rebuilt by his neighbour, Paddy Umney, during the 1970s. The keel and some treenails were the only original parts of the boat remaining when he had finished. Welland put a mast and sails on her as well as an engine. In 1984, she was bought by Thomas Nash of Cobh. Her next owner was John Hunt of Bantry who changed her name to the *Karen Marie*. He sold her to Martin Bennett who in turn sold her to Finbarr Goggin, both from Bantry. During her years in Bantry, a wheel-house has been added. She can frequently be seen tied up at the old quay in Bantry.

Three stages of a metamorphosis. First, the *Mary Ellen* (*above*) photographed in July 1932. On board (*left to right*) are Jer McCarthy, Patsy Murphy, Danny Harte and her owner, Danny O'Neill, all from Heir Island. Second (*left*), the *Two Fishes* (formerly the *Mary Ellen*) in the late 1970s after she had been rebuilt by Tony Welland and Paddy Umney of Monkstown. Finally (*below*), the *Karen Marie* (formerly the *Two Fishes*) in Bantry in 1998, with a wheel-house.
Photograph below: Cormac Levis

Valetta S852

Builder:	Richard Pyburn, Heir Island	
Date built:	1902	
Date registered:	9/2/1903	
Registered dimensions:	Length of keel:	26.0 ft
	Length overall:	27.1 ft.
	Breadth:	7.0 ft.
	Depth:	3.2 ft.
	Tonnage:	3.81
Rig:	Gaff mainsail, foresail and jib	

The *Valetta* was built for Dan Murphy of Heir Island but was fished by John O'Donovan, also of Heir Island. In 1925, John O'Donovan became her owner, and she later passed on to John's son, also named John. She is said to have been a good sailer. The *Valetta* was probably named after the SS *Valetta* which was part-owned by the Kelsall Brothers of Liverpool who were among the main fish merchants buying mackerel at Baltimore in the 1890s. In 1962, the *Valetta* was sold to someone who had returned from the US to live in Baltimore. In 1964, she was bought by Donal Dinan of Cork. By this time, she had been fitted with a 6/7 h.p. Kelvin engine and a foredeck, and her gunnels had been raised. The *Valetta* was by now in poor condition, and Dinan rebuilt her over a period of six years, adding a wheel-house. He fished her for lobsters out of Sherkin Island for some years before she broke away from a mooring in Baltimore Harbour and was holed and sank at Baltimore pier. The *Valetta* was repaired at Skinner's of Baltimore and a 16 h.p. Volvo diesel engine fitted. Donal Dinan sold the *Valetta* to his son Paul who later sold her to Newell Cave of Blackrock, Cork. Having spent some years hauled up on the slip at Blackrock, she was eventually sold on to Peter Cullen of Ballintemple who intended to restore her but eventually laid her up in George Hegarty's yard in Rafeen, Monkstown (*see photograph* p. 17).

Victory II S14 also S630

Builder:	Unknown, but built in Baltimore	
Date built:	1910	
Rig:	Gaff mainsail and foresail	
S630		
Date registered:	14/6/1910	
Registered dimensions:	Length overall:	26.4 ft.
	Breadth:	6.8 ft.
	Depth:	3.2 ft.
	Tonnage:	3.35
S14		
Date registered:	17/4/1929	
Registered dimensions:	Length of keel:	23.0 ft.
	Length overall:	26.0 ft.
	Breadth:	6.8 ft.
	Depth:	4.5 ft.
	Tonnage:	3.58

This boat was first registered on the 14/6/1910 under the number S630. This registration was closed on 31/12/1926 when it was stated that she 'ceased to be a fishing boat'. She was re-registered, however, under the number S14 on the 17/4/1929. The differences between the two registrations as regards overall length, depth and tonnage would indicate that she had undergone some remodelling by the time

Victory II picking up passengers at Whitehall for Baltimore Regatta in the late 1930s. The two men wearing hats, one at the stern and one at the stem, are her owner Michael Coughlan of Kilkilleen, Aughadown and Patie Hegarty, Whitehall. In the foreground are their sons, John Coughlan and Jeremiah Hegarty. (See also photograph on p. 52.)

she was registered for a second time. Alternatively, it is possible that errors were made in her length and depth measurements, which in turn led to a discrepancy between the tonnages recorded.

Victory II was originally owned by Michael Coughlan of Kilkilleen, Aughadown. She left the fleet around 1940 when she was sold to Patrick Cahalane of Droumbeg, Glandore. In 1948, she was bought by Alexandra Sheele of Coolmain, Kilbrittain who changed her name to the *Coolmain*. Her registration was closed in 1951.

Village Maid S889

Builder:	Richard Pyburn, Heir Island	
Date built:	1896	
Date registered:	10/6/1914	
Registered dimensions:	Length overall:	24.0 ft.
	Breadth:	6.9 ft
	Depth:	3.0 ft.
	Tonnage:	3.24
Rig:	Gaff mainsail and foresail	

The *Village Maid* was registered in the name of John Harte, Heir Island. The last entry in the *Register* is dated 11/2/1925. This is probably John Harte who had another boat, the *St Margaret* S47, built in that year.

Water Lily S701 also S186

Builder:	The Pyburns, Toormore	
Date built:	1904	
Rig:	Gaff mainsail and foresail	
S701		
Date registered:	25/6/1908	
Registered dimensions:	Length overall:	26.0 ft.
	Breadth:	6.9 ft
	Depth:	2.6 ft.
	Tonnage:	3.31
S186		
Date registered:	13/7/1932	
Registered dimensions:	Length overall:	26.0 ft.
	Breadth:	6.6 ft.
	Depth:	3.0 ft.
	Tonnage:	2.32

The *Water Lily* was originally owned by Johnny 'Florrie' O'Driscoll of Long Island who was drowned when she was capsized by a *sí gaoithe* in Long Island Channel in 1908. She was inherited by his son Jack, who continued to fish her for lobsters until he sold her to Michael McCarthy of Heir Island in whose name she was re-registered in 1932. There is an unexplained discrepancy between the two registrations as regards the breadth, depth and tonnage measurements. McCarthy fished her for some years before breaking her up in 1941.

***White Lily* S635**

Builder:	Unknown, but built at Baltimore	
Date built:	1910	
Date registered:	27/6/1910	
Registered dimensions:	Length of keel:	22.5 ft.
	Length overall:	27.6 ft
	Breadth:	7.0 ft.
	Depth:	3.2 ft.
	Tonnage:	3.66
Rig:	Gaff mainsail and foresail	

The *White Lily* was originally owned by Michael McCarthy of Lisaree, Aughadown and later by his son Pat. Around 1941, she was bought by Jim O'Sullivan of Skeaghanore, Ballydehob who used her as a sand-boat. She was later owned by Jim's brother Miah, who put her to the same use. She was broken up at Skeaghanore in the 1950s. She had never been fitted with an engine.

White Yawl (see ***Lily* S125**)

Lobsterboat owners, the names of whose boats have yet to be ascertained:

Tim Desmond, East Skeam Island.

Donal O'Neill, Heir Island.

Will Morgan, Fasagh, Aughadown.

Tom 'John' Whooley, Long Island.
His boat was possibly known as the *Eileen*.

Paddy 'Mick' Regan and his son Mike, Long Island.

Mike 'Donnchadh' O'Donovan, Long Island.
His boat went out of commission in the early 1900s.

Tim and Patrick Murphy, Heir Island.
Theirs was one of the boats that had to be hauled up at Ardmore during a gale early in the twentieth century.

Mike 'the Cloud' O'Neill, Heir Island.
The *Cloud* is said to have been the name of a large, Manx mackerel drifter he once owned. However, in the 1901 census returns, in a file covering shipping in Heir Island and Collatroum, *Primrose* is the name given for a thirty-three-ton, Manx-registered fishing boat, owned by one Michael Neill. If the *Primrose* was, indeed, Mike 'the Cloud's' mackerel drifter, then it is possible that the *Cloud* (or *Flying Cloud*) was the name of his lobsterboat. In any event, his lobsterboat was eventually sold to someone in Dunmore East, Co. Waterford.

Tom Fuller, Skibbereen.
Fuller was a businessman who owned a lobsterboat and employed Jerry McCarthy and other Heir Islanders to fish her up to the 1920s. She had an engine installed at a very early stage. Jerry McCarthy, John Harte and Tim Murphy had been herring fishing in this boat on the night the *Thomas Joseph* was lost (*see* p. 77). On hearing cries for help, they first located the *Thomas Joseph* and then picked up a smaller boat on Heir Island with which Harte carried out his heroic rescue. It is quite likely that it was with Fuller's boat, also, that the Hartes and Cahalanes went to the rescue of the *Saurnaut* in 1916. If that was indeed the case, then her name was the *Mary Anne*.

Notes

Abbreviations

CDB Congested Districts Board for Ireland
IFC Irish Folklore Collection,
Roinn Bhéaloideas Éireann/Department of Irish Folklore,
University College, Dublin

1. *Annual Report of the Inspectors of Irish Fisheries on the Sea and Inland Fisheries of Ireland, 1881*, p. 6, cited in Seamus Fitzgerald, *Mackerel and the Making of Baltimore, County Cork 1879–1913*, Maynooth Studies in Local History, no. 22 (Dublin and Portland, OR, 1999), p. 15.
2. CDB, *Baseline Reports of Local Inspectors*, 1892:
   No. 78 District of Kenmare, p. 3;
   No. 79 District of Castletown Bearhaven, p. 3;
   No. 81 District of Sheepshead, p. 2;
   No. 82 District of Skull, p. 3;
   No. 83 District of Baltimore, p. 2;
   No. 84 District of Courtmacsherry, p. 2.
3. Donncha Ó Cróinín (ed.), *Seanachas ó Chairbre* 1 (Baile Átha Cliath, 1985), p. 178.
4. The CDB was established under the 1891 Land Act to improve living conditions for the inhabitants of some of the poorest parts of Ireland.
5. The CDB *Baseline Reports* for the District of Baltimore (p. 2.) confirms that the number of men employed on the lobsterboats in the electoral district of Aughadown South was 120, i.e. three per boat. In all other areas where any lobsterboats were based, including the Schull District, the number given was four times that of the number of boats.
6. In 1901, Long Island (352 acres) had thirty-six families and a population of 199. West Calf (sixty-two acres) had three families and a population of twenty-seven. Middle Calf (sixty-four acres) had four families and a population of eighteen.
7. IFC, vol. 1320, p. 486.
8. A beam fixed to the inside of the stem-post, from the deadwood up, to which the forward ends of the planks are fastened.

9. A word derived from the Irish word *stiúir*, meaning a rudder.
10. A beam running along the top of the keel from deadwood to deadwood. It passes beneath the floors and would facilitate the easy replacement of the keel in the event of it being badly damaged.
11. A tool like an axe but having an arched blade at right angles to its handle.
12. These legwales or bends were also a feature of the Ballydehob sandboats, and the legs were attached in exactly the same way, except that in the sandboats, an iron 'box' from the axle of a horse cart was forced on over the bottom of the leg to ballast it and ensure that it remained in a vertical position when the boat floated. The notched frames and overlapping sheer strake were also a feature of the Ballydehob sandboats. See Cormac Levis, 'The Almost Forgotten Harbour', *Mizen Journal*, vol. 4 (Schull, 1996).
13. The timbers of a grown oak frame are cut from oak that has a natural bend in it corresponding with the shape of the frame. Thus, the grain of the wood runs the length of the timbers, giving a strong frame, and the entire frame can be made up of three timbers, i.e. one top timber at either side and a ground timber or floor. If it is not possible to procure grown oak, i.e. oak with suitable bends, then a shorter top timber is used at either side and a middle timber, or futtock, is introduced between it and the floor. The top timber is butted against the futtock and held in position by another length of oak, known as a butt strap, which is cut to the shape of the frame and bolted to the side of the futtock and the side of the top timber. Alternatively, the top timber and futtock are allowed to overlap a certain distance and simply bolted side to side. Both methods effectively double the width of this part of the frame.
14. Presumably derived from the word *tochta*, the Irish for thwart.
15. IFC, vol. 1320, p. 487.
16. From a letter written by Richard Desmond, originally from the East Skeam Island, and transcribed in Catherine Dewsnap (Desmond), 'Emigrant Islanders Correspond', in Brigid O'Brien and Mary Whooley (eds.), *From Ilen to Roaringwater Bay – Reminiscences from the Parish of Aughadown* (Aughadown, 1999), p. 112. Richard Desmond fished on the lobsterboats before he emigrated to the US around 1921.
17. IFC, vol. 1517, p. 50.
18. A tanbark rich in tannin and bruised or cut into small pieces. It could be bought in local shops. The bark of various species of oak was frequently used. Burma cutch, a dry extract from the wood of the tropical acacia catechu, and Borneo cutch, from the bark of a mangrove, were also commonly used.

19. 1 nautical mile = 1.15 statute miles/1.85 k.m.
20. IFC, vol. 1517, p. 70.
21. IFC, vol. 742, p. 405.
22. The reesing, or rising, is a stringer or strengthening beam running the length of the hull at either side of the boat.
23. IFC, vol. 1320, p. 440; this local spelling of the word does not seem to appear in any dictionary of the Irish language, but the following entry is found in Rev. Patrick S. Dinneen, *Foclóir Gaedhilge agus Béarla* (Dublin, 1927, reprinted 1970), p. 1151: 'tabhall . . . a tablet, plate or slab, a balk, a foot-plank, a roof or loft; al. a sling'. An tUasal Cian Ó Sé has very kindly sourced the following references to similar words:
teamhal: 'Covering to keep rain off boat', in L. Mac Cionnaith SJ, *Foclóir Béarla agus Gaeilge* (Oifig Díolta Foillseacháin Rialtais, 1935), p. 259;
téalta: 'm., An improvised hut or tent (nautical), a shelter', in Rev. P.S. Dinneen, op. cit., p. 1189;
shelter: (3) Teallta. 'Make a shelter of this big sail, Déan teallta den tseol mór seo', in *Lane's Larger English-Irish Dictionary* (Talbot Press, 1922), p. 1399.
The words *téalta* and *teallta* above are of Old Norse origin which is closely related to Icelandic, and the following entry is found in Geir T. Zoëga, *Concise Dictionary of Old Icelandic* (Oxford, 1910, latest edn. 1975): *tjald*: 'n. (1) Tent, on land or on ships, esp. when in harbour'.
24. IFC, vol. 1320, p. 493.
25. Ibid., p. 441.
26. In Roaringwater Bay, the term 'yawl' was a very general one, referring to any boat between approximately eighteen feet and thirty-three feet in length, regardless of rig.
27. An oven-pot. The bastible was normally used for baking bread on an open fire. It had a lid and was covered over with live turves or coals.
28. A small cove or creek.
29. During World War II, when there was a shortage of oil and candles, people reverted to making their own candles once more. Saving the candle grease from lighting candles, they re-melted it and poured it into a mould of brass piping. Bottles were also used as moulds. A crocheted, wool wick was used, the diameter of which depended on the diameter of the mould. Great expertise was required to produce a wick which was thick enough not to be quenched by the melting candle grease and thin enough to avoid melting the candle too quickly.
30. A light seaweed.
31. A nephew of the famous John Harte whose exploits are detailed in ch. 15, 'Illustrious Lobstermen'. He was no immediate relative of his skipper, Dan Harte.

32. An area in the farmyard where ricks of hay or stacks of corn were built.
33. Cove of the Water.
34. Presumably from the word 'wash-beetle', a stick or club used for beating or pounding clothes when washing.
35. *Gréasaí* literally means 'shoemaker'.
36. *Burd* is presumably an abbreviation of *burdán* which means 'a tiny fellow'. Hence, *Burd Buí* means 'a tiny yellow fellow'.
37. Withe, withy, 'n. Tough flexible branch esp. of willow or osier used for binding bundles etc.': J.B. Sykes (ed.), *The Concise Oxford Dictionary of Current English* (Oxford, 6th edn., 1976), p. 1340.
38. This illustration is based on detailed descriptions supplied by several informants, two of whom were pot-makers in their younger days. An account of pot-making submitted to the Department of Irish Folklore, UCD in 1940 by Ciarán Ó Síothcháin of Cape Clear describes a mould with a much larger disc, two and a half feet in diameter, with the circle of holes at its centre (IFC, vol. 742, p. 399). Ó Síothcháin notes that there were a number of different types of mould in use.
39. In the Irish of Cape Clear, the standards were referred to as *saighti* (singular: *sagha*) and the main *buinne* was known as '*buinne na saighti*'. The *buinne* around the bottom edge of the pot was known as '*buinne na tóna*', meaning the 'wale of the bottom'. IFC, vol. 742, pp. 400–3.

    Fig. 3 and fig. 4 on p. 58, and the photograph on p. 45 show the main *buinne* spiralling down around the pot in a clockwise direction and the twigs of the *buinne* twisted anti-clockwise around each other. In actual fact, it was more often the case that the *buinne* spiralled in an anti-clockwise direction while the twigs were twisted clockwise, as this was the natural inclination of a right-handed pot-maker.
40. F.M. Davis, *An Account of the Fishing Gear of England and Wales*, Ministry of Agriculture, Fisheries and Food Fishery Investigations, series II, vol. xxi, no. 8 (London, 4th edn., 1958, (first published 1923)), p. 126.
41. Twelve and a half pence or sixteen cent.
42. Pádraig Ó Maidín, 'Pococke's Tour of South and South-West Ireland in 1758', *Journal of the Cork Historical and Archaeological Society*, lxiii (1958), p. 82.
43. Wallop Brabazon, *The Deep Sea and Coast Fisheries of Ireland, with Suggestions for the Working of a Fishing Company* (Dublin, 1848), p. 67.
44. Nets with large meshes in which crayfish, lobsters, crabs and large flat-fish are easily entangled.
45. Robert Wilson, jnr., *A Brief Description of Cape Clear* (Cork, 1832), extract published in James Coombes (ed.), *Seanchas Chairbre* (Skibbereen, 1983), p. 28.

46. Wallop Brabazon, op. cit., p. 67.
47. Ibid., p. 67.
48. *First Report of the Commissioners of Inquiry into the State of the Irish Fisheries* (henceforth cited as 1837 *Inquiry* report), HC 1837 (77), xxii,1, p. 151.
49. Ibid., p. 152.
50. Ibid., p. 152.
51. Wallop Brabazon, op.cit., p. 68.
52. 1837 *Inquiry* report, p. 152.
53. Ibid., p. 156.
54. Ibid., p. 156.
55. Ibid., p. 156.
56. *Report on the Coast and Deep Sea Fisheries of Ireland, by the Royal Commissioners on Irish Oyster Fisheries*, HC 1870 (c. 226), xiv, 231, pp. 12–13.
57. F.M. Davis, op. cit., p. 124.
58. Department of Agriculture and Technical Instruction for Ireland, *Ireland Industrial and Agricultural* (Dublin, Cork and Belfast, 1902), p. 373.
59. Seamus Fitzgerald, op. cit., p. 15.
60. Ibid., p. 30.
61. CDB, *Baseline Reports of Local Inspectors*, 1892, no. 83 District of Baltimore, p. 2.
62. A two-masted vessel with a gaff mainsail and a standing lug mizen.
63. *Cork Examiner*, 27 April 1894.
64. *Cork Examiner*, 26 April 1894.
65. *Cork Examiner*, 26 April 1894.
66. *Cork Examiner*, 17 April 1894.
67. A wool-like seaweed used as a fertiliser.
68. Little Cove of the Corpses.
69. From the Irish *feileastram* – wild iris.
70. The RNLI citation presented to John Harte in the possession of his daughter-in-law, Mrs Breda Harte, Laharantseanvally, Aughadown.
71. *Eagle and County Cork Advertiser*, 30 April 1892. In this brief report on the outcome of the trial, the boy's name is given as Walker. In the *Southern Star* of 20 Feb. 1892, the name of the schooner is given as the *Petro*.
72. *Southern Star*, 28 May 1892. Remarkably, while a list of the names of the dignitaries present at the ceremony is given in this report, the recipients of the awards are not referred to by name at all.
73. Dulse, an edible seaweed.
74. *Eagle and County Cork Advertiser*, 15 Aug. 1896.
75. *Eagle and County Cork Advertiser*, 10 Oct. 1896.
76. All of this information regarding lobster cutters at Kinsale,

including the quotations, is from the Kinsale Harbour-Master's Book of Arrivals and Sailings.

77. The *Smelt* had a ketch rig but also apparently had a 30 h.p. steam engine, as recorded in the 'Alphabetical List of British Registered Steam Vessels' in J. Blake Harrold, Registrar-General of Shipping and Seamen, *The Mercantile Navy List and Maritime Directory for 1924* (London). She was fifty-seven and a half feet in length and was forty-one tons gross. She was built at Grimsby in 1892 and registered in Cork in 1911 with Keane as her managing owner.
78. Seoirse Mac Tomáis (eagarthóir), Muiris Ó Súilleabháin, *Fiche Bliain Ag Fás* (An Daingean, 1998), p. 150.
79. 'Alphabetical List of British Registered Motor Vessels', in J. Blake Harrold, op. cit. The *Mary Leak* was almost fifty-eight feet in length and was forty-two tons gross. She was built in Grimsby in 1893 and registered in Cork in 1917 with Saunders listed as her owner and Scott as her manager.
80. A term used in reference to a type of fishing boat built at Buckie in north-east Scotland. Many of these boats were imported, second-hand, into Ireland.
81. *Southern Star*, 4 Aug. 1928.
82. The *Eugenie* was built in Roscoff in 1905, the *Edmond* in Paimpol in 1912 and the *Trimardeur* in Camaret in 1905.
83. The *Walblow* was eventually wrecked at Fenit, Co. Kerry.
84. The *Ariel* was built in 1928 and the *Notre Dame d'Esperance* in 1932; both were built in Roscoff. The *Fleur de France* was built in Granville in 1934.
85. Jack Pyburn's son Charles continued to repair boats at Crookhaven down to about 1950. His other son, Tom, spent most of his life at sea as a ship's carpenter, like his progenitor centuries earlier.
86. Willie Skinner's yard at Rathmore closed down in the early 1950s.